W0245994

81 Structure and Bonding

Editors:
M. J. Clarke, Chestnut Hill, MA
J. B. Goodenough, Austin, TX • J. A. Ibers, Evanstone, IL
C. K. Jørgensen, Genève • D. M. P. Mingos, London
J. B. Neilands, Berkeley, CA • G. A. Palmer, Houston, TX
D. Reinen, Marburg • P. J. Sadler, London
R. Weiss, Strasbourg • R. J. P. Williams, Oxford

Structures and Biological Effects

With contributions by
R. E. Benfield V. Chandrasekhar M. J. Clarke
M. W. Dirken J. B. Gaul S. Mazumdar S.Mitra
H. H. A. Smit R. C. Thiel K. R. Justin Thomas
R. Zanoni

With 48 Figures and 15 Tables

Springer-Verlag Berlin Heidelberg GmbH

ISBN 978-3-662-14925-6   ISBN 978-3-540-47567-5 (eBook)
DOI 10.1007/978-3-540-47567-5

© Springer-Verlag Berlin Heidelberg 1993
Originally published by Springer- Verlag Berlin Heidelberg New York in 1993
Softcover reprint of the hardcover 1st edition 1993

Typesetting: Macmillan India Ltd., Bangalore-25, India
51/3020 - 5 4 3 2 1 0 - Printed on acid-free paper

Editorial Board

Attention all "Structure and Bonding" readers:

A file with the complete volume indexes Vols. 1 through 81 in delimited ASCII format is available for downloading at no charge from the Springer EARN mailbox. Delimated ASCII format can be imported into most databanks.

The file has been compressed using the popular shareware program "PKZIP" (Trademark of PK ware Inc., PKZIP is a available from most BBS and shareware Distributors).

This file is distributed without any expressed or implied warranty.

To receive this file send an e-mail message to:

SVSERV@DHDSPRI6.BITNET.

The message must be: "GET /CHEMISTRY/SB_V1.ZIP".

SVSERV is an automatic data distribution system. It responds to your message. The following commands are available:

HELP	returns a detailed instruction set for the use of SVSERV,
DIR (*name*)	returns a list of files available in the directory "name",
INDEX (*name*)	same as "DIR",
CD <*name*>	changes to directory "name",
SEND <*filename*>	invokes a message with the file "filename",
GET <*filename*>	same as "SEND".

Table of Contents

The Physical Properties of the Metal Cluster
Compound $Au_{55} (PPh_3)_{12}Cl_6$
R. C. Thiel, R. E. Benfield, R. Zanoni, H. H. A. Smit,
M. W. Dirken 1

Recent Aspects of the Structure and Reactivity
of Cyclophosphazenes
V. Chandrasekhar, K. R. Justin Thomas 41

Biomimetic Chemistry of Hemes Inside Aqueous Micelles
S. Mazumdar, S. Mitra 115

Chemistry Relevant to the Biological Effects
of Nitric Oxide and Metallonitrosyls
M. J. Clarke, J. B. Gaul 147

Author Index Volumes 1 - 81 183

The Physical Properties of the Metal Cluster Compound $Au_{55}(PPh_3)_{12}Cl_6$

R. C. Thiel[1], R. E. Benfield[2], R. Zanoni[3], H. H. A. Smit[1,4] and M. W. Dirken[1,4]

[1] Kamerlingh Onnes Laboratory, Leiden University, P.O. Box 9506, 2300 RA, Leiden, The Netherlands
[2] Chemical Laboratory, University of Kent at Canterbury, Kent CT2 7NH, United Kingdom
[3] Dipartimento di Chimica, Università Degli Studi di Roma "La Sapienza", Ple. A. Moro, 5, 00185 Rome, Italy
[4] Present address: Philips I & E, Almelo, The Netherlands

A discussion is given of the thermal, electronic, and magnetic properties of the gold cluster compound $Au_{55}(PPh_3)_{12}Cl_6$. These results are combined to form a consistent physical picture of the system. The coordination numbers of the individual gold sites in this cluster compound play an essential role.

It is shown, making use of the results of new physical measurements as well as information from the physical and chemical literature, that the gold core of this cluster material already exhibits metallic bonding, while at the same time displaying some of the characteristics of a discrete energy level spectrum.

1 Introduction 2

  The Structure of Au_{55} 4
  2.1 A Sketch of the Structure of Au_{55} 4
  2.2 EXAFS Measurements 6
  2.3 Differential Scanning Calorimetry Results 6

3 The Thermal Behavior of Au_{55} 8
  3.1 The Thermal Behavior at Low Temperature 8
  3.2 A Model Description of the Thermal Behavior at Low Temperature 10
  3.3 The Low Temperature Specific Heat of Au_{55} 11
  3.4 A Surface Mode Description in Relation to the MES results 12
  3.5 The Melting Temperature of Au_{55} 12
  3.6 Sample Preparation 13

4 The Electronic Structure of Au_{55} 14
  4.1 Electrons in a Small Particle of Gold 14
  4.2 The Metallic Binding of the Cluster Core 17
  4.3 The Isomer Shift of the 13 Core Sites of Au_{55} 18
  4.4 The Metallic Binding of the Surface Sites 19
  4.5 Other Experimental Evidence for Metallic Binding 20
  4.6 ESR Measurements on Au_{55} 23
  4.7 X-Ray and UV-Visible Absorption on Au_{55} 24

5 Photoelectron Spectroscopy on Au_{55} 26
  5.1 XPS Results on Au_{55} 26
  5.2 Analysis of the Au4f Level 29
  5.3 The Valence Band Results Compared to XPS on Other Systems 31
  5.4 Comparison of XPS to MES and EXAFS Results 32
6 Conclusions 33

7 References 35

1 Introduction

There has been an increasing interest during the past decade in the physical study of metal clusters in the form of large molecules, with the metal cluster core surrounded by a number of ligands [1]–[23]. Such cluster compounds, with a well defined size and shape, are available in macroscopic quantities. This has made it possible to study the properties of individual clusters, and in some cases even of individual sites within such clusters [24, 25].

Of the larger metal cluster compounds, probably the most studied has been the compound $Au_{55}(PPh_3)_{12}Cl_6$ (where $PPh_3 = Ph_3P = P(C_6H_5)_3$) [26], with the largest metal core in the series of gold cluster compounds known to date. Two related compounds recently reported, $Au_{55}[Ph_2PC_6H_4SO_3Na]_{12}Cl_6$, which is a water soluble derivative [27], and $Au_{55}[P(C_6H_4CH_3)_3]_{12}Cl_6$, which is a methylated derivative [28], also have cores of 55 gold atoms. 55-atom clusters of Rh, Ru, and Pt have been prepared as well [31, 16, 32].

$Au_{55}(PPh_3)_{12}Cl_6$ is an amorphous material, and it has not yet proven possible to grow crystals suitable for direct structure determination by X-ray crystallographic analysis. A cuboctahedral 55-atom cluster structure (Fig. 1) was originally proposed from ^{197}Au Mössbauer measurements [26]. Subsequently, there has been some controversy about whether $Au_{55}(PPh_3)_{12}Cl_6$ is a pure compound or a colloidal mixture of several components. Fackler et al. [33] reported that ^{252}Cf plasma desorption mass spectra of several independently prepared samples of $Au_{55}(PPh_3)_{12}Cl_6$ showed the presence of equilibrium mixtures of clusters in three broad mass ranges. The major component was formulated as $Au_{67}(PPh_3)_{14}Cl_8$, and it was suggested [33] that it has a polyicosahedral structure of the 'supracluster' type known from the work of Teo [34]. However, subsequent measurements by Feld et al. [35, 36] using secondary ion mass spectroscopy have challenged this interpretation, giving fragmentation patterns consistent with a cubic close-packed Au_{55} cluster. Teo himself has commented [37] that a polyicosahedral Au_{67} structure is unlikely to occur without severe steric distortion.

All the other available experimental structural and spectroscopic evidence on $Au_{55}(PPh_3)_{12}Cl_6$, which we discuss in detail in this paper, supports a cuboctahedral 55-atom structure for the gold clusters in this material; most of it is inconsistent with icosahedral structures or the presence of a mixture of clusters. We also note that $[Au_{39}(PPh_3)_{14}Cl_6]^{2+}$, the largest cluster compound of gold which has been crystallographically characterised to date [38], has a structure related to a hexagonal close-packed geometry, not an icosahedral one. We consider the properties and bonding of the gold clusters in $Au_{55}(PPh_3)_{12}Cl_6$ in this context.

In this paper, only the original triphenyl phosphine compound $Au_{55}(PPh_3)_{12}Cl_6$ will be examined. In order to simplify reading, we will use the abbreviation Au_{55} to mean this version. For an occasional reference to the water soluble version, this will be marked as Au^*_{55}, and the methylated version will be noted as Au_{55}-tolyl. Related smaller centered clusters, such as $[Au_9L_8]^{3+}$ [4] and $Au_{11}L_7X_3$ [5, 6], will be used occasionally as reference compounds. Throughout this paper $L = PPh_3$ or a similar phosphine ligand and X = a halide or $[NCS]^-$ ion; otherwise the cluster formulae will be labeled specifically.

Thanks to the extensive literature on Au_{55} and the related smaller gold cluster compounds, plus some new results and reanalysis of older results to be presented here, it is now possible to paint a fairly consistent physical picture of the Au_{55} cluster system. To this end, the results of several microscopic techniques, such as Extended X-ray Absorption Fine Structure (EXAFS) [39, 40, 41], Mössbauer Effect Spectroscopy (MES) [24, 25, 42, 43, 44, 45, 46], Secondary Ion Mass Spectrometry (SIMS) [35, 36], Photoemission Spectroscopy (XPS and UPS) [47, 48, 49], nuclear magnetic resonance (NMR) [29, 50, 51], and electron spin resonance (ESR) [17, 52, 53, 54] will be combined with the results of several macroscopic techniques, such as Specific Heat (Cv) [25, 54, 55, 56, 49], Differential Scanning Calorimetry (DSC) [57], Thermogravimetric Analysis (TGA) [58], UV-visible absorption spectroscopy [40, 57, 17, 59, 60], AC and DC Electrical Conductivity [29, 61, 62, 63, 30] and Magnetic Susceptibility [64, 53]. This is the first metal cluster system that has been subjected to such a comprehensive examination.

In Sect. 2, the structure of Au_{55} is discussed, utilizing the interatomic distances obtained from EXAFS [39, 40, 41]. DSC results [57] give further evidence that Au_{55} is a stoichiometric molecular compound, and recent SIMS results [35, 36], confirm the molecular mass, the structure, and the stoichiometry of this compound, as originally postulated by Schmid [16].

The thermal properties of Au_{55} are treated in Sect. 3, using especially the results of MES measurements [24, 25, 42]. These are discussed in connection with the concept of bulk versus surface modes in small particles. An explanation of the temperature dependence of the MES [42] absorption intensities and the Cv results [25] on the basis of a model using the site coordination and the center-of-mass motion are briefly reviewed. The consequences of the Mössbauer results for surface Debye temperatures and for the melting temperature of small gold particles are also discussed.

In Sect. 4, some aspects of the electronic structure of small gold particles in general, and of Au_{55} in particular, are considered. The question of metallic bonding of the gold atoms in the cluster is discussed, utilizing the reported results of a number of different measurement techniques.

The model of Citrin and Wertheim [65], developed to explain the surface atom core level shifts in the XPS spectra of metallic systems, has already been used to explain the extremely low value of the I.S. of the central 13 atoms in the Au_{55} cluster [44].

The X-ray absorption edge of Au_{55} from EXAFS, as compared to that of bulk gold [39, 40, 41], also suggests that the entire metal core of this compound exhibits metallic bonding.

It is shown that the semi-empirical model of Miedema and v.d. Woude [66], which was developed for predicting the I.S. changes of Mössbauer nuclei in alloys and intermetallic compounds, gives a remarkably good prediction of the observed I.S. values for the surface sites [67].

AC and DC electrical conductivity results [29, 61, 62, 63, 30] are also consistent with the picture of metallic bonding. To give an adequate description of these conductivity results it is necessary to take into consideration the inter-cluster interactions, which slightly modify the electronic structure of the individual clusters. For understanding these results, a discrete energy level spectrum seems to be necessary. ^{31}P NMR [29, 50, 51] gives information related especially to the inter-cluster interactions. Magnetic susceptibility as a function of temperature and field [64, 53], on the other hand, also indicate that there is probably a discrete electronic energy level spectrum in the cluster system, and not a true continuum. The lack of a linear term in the low temperature specific heat [25, 54, 56] is discussed.

Although ESR measurements on Au_{55} have been attempted [17, 52, 53, 54], the results remain ambiguous. The situation will be discussed briefly and a comparison made to ESR results on gold colloids and gold foil.

Kreibig et al. [68] have recently proposed three alternatives for understanding the lack of a well developed plasma resonance in the UV-visible absorption spectra of Au_{55} [40, 57, 17, 59, 60] which one would normally expect with small metallic particles of gold [69, 70, 71, 72, 73]. These are examined here, together with a fourth possible explanation.

New XPS data on Au_{55} [48, 49] are presented in Sect. 5. There appear to be serious discrepancies between these results and XPS/UPS data recently published elsewhere [47]; these discrepancies are discussed and accounted for.

We also compare our results to XPS data on bulk gold and on the smaller centered cluster compound $Au_{11}L_7X_3$ (with X = Cl or I) [74]. A comparison is also made to XPS results obtained on bare gold clusters deposited on poorly conducting substrates [75, 76]. This gives still more support to the idea of the metallic bonding of the Au_{55} clusters.

Comparison of these XPS results and their correlation to the MES I.S. results [24] is discussed.

A short summary and a discussion of the question of where the boundary between molecular and metallic behavior lies for gold clusters and cluster compounds are given in Sect. 6.

2   The Structure of Au_{55}

2.1   A Sketch of the Structure of Au_{55}

The results of a number of recent microscopic measurement techniques [24, 39, 40, 41] and model calculations [25] can be combined to define the structure of Au_{55}. The assumed structure will first be described, in order to simplify the discussion.

The structure of the gold core, shown schematically in Fig. 1 without the surrounding ligands [16], has three shells of gold atoms in an fcc stacking, comprising 1, 12, and 42 atoms per shell.

The gold atoms comprising the two inner shells of this cuboctahedral arrangement are fully coordinated by twelve gold nearest neighbors. On the cluster surface are 12 gold vertex atoms, with 5 gold nearest neighbors and ligated by the phosphorus atoms of the PPh_3 (triphenyl phosphine) groups. There are 6 gold atoms at the centers of the square faces, each having eight gold nearest neighbors and ligated by chlorine. And there are 24 bare (i.e. unligated) gold edge atoms, each coordinated by seven gold nearest neighbors and sterically screened by the ligands on neighbouring surface gold atoms. These unligated gold atoms on the cluster surface are noteworthy; such atoms are not found in small (2-shell) gold clusters [15], but are known in the larger $[Au_{39}(PPh_3)_{14}Cl_6]^{2+}$ [38]. The "diameter" of the three shell gold core is about 1.4 nm. See also Table 1.

Since Au_{55} precipitates out of solution in the form of a fine amorphous powder, structure determination on the basis of single-crystal X-ray diffraction is not possible [16]. On the other hand, high resolution electron microscopy [77] has produced photographs which strongly resemble the structure outlined above. Scanning tunneling microscopy [78, 79, 80] is also consistent with this structure.

Fig. 1. The surface sites of the Au-core of Au_{55} are shown schematically, with the ligands removed. The *light grey circles* represent the uncoordinated surface atoms, the *dark grey circles* represent PPh_3-coordinated surface atoms, and the *black circles* represent the Cl-coordinated surface atoms. The 13 atoms of the two inner shells are not visible

Table 1. The sites of Au_{55}, the site coordination numbers, the model values of the I.S., including a correction mentioned in the text, and the difference between model and experimental values. I.S. values are relative to Au in Pt at 4.2 K.

	site occupation	coörd. no.	model value I.S. (mm/s)	expl. value I.S. (mm/s)	difference (I.S.)
bulk	—	12	—	−1.2	—
core	13	12	ref. value →	−1.4	—
Cl^-	6	8	−0.4	+0.1	−0.5
bare	24	7	0.0	+0.3	−0.3
PPh_3^-	12	5	+1.0	+0.6	+0.4

2.2   EXAFS Measurements

EXAFS measurements on Au_{55} prepared by Schmid [39, 41], as well as on independently prepared samples [40], have shown clearly that there is only a single interatomic distance between the gold atoms in this material, a distance about 4% shorter than that in bulk gold.

An icosahedron has shorter radial than tangential interatomic distances [81]. In the icosahedral $[Au_{13}(PMe_2Ph)_{10}Cl_2]^{3+}$ cation [82], the average radial gold-gold distance is the same as that for the inter-gold distance in Au_{55}, but the average tangential gold-gold distance is nearly 5% larger. The same effect had been found in the icosahedral $[Au_{13}(dppm)_6](NO_3)_4$ [1, 8], but in this case crystallographic disorder limited the accuracy of the bond length measurements.

EXAFS measurements on the $Au_{11}[PPh_2(p\text{-}ClC_6H_4)]_7I_3$ cluster compound [41], whose structure is based upon an incomplete icosahedron [7], clearly resolve two different inter-atomic distances between gold atoms, in agreement with the results of single crystal X-ray crystallographic measurements on this type of Au_{11} cluster [5, 6, 9]. Thus, we have proven that EXAFS can be used to distinguish between cubic close-packed and icosahedral gold cluster structures.

The single interatomic distance found for the Au_{55} compound, together with the comparison to $Au_{11}[PPh_2(p\text{-}ClC_6H_4)]_7I_3$ where multiple inter-gold distances could indeed be measured, constitutes clear proof that Au_{55} has a cubic close-packed arrangement, not an icosahedral or polyicosahedral structure as proposed by Fackler [33]. There are two basic arguments for expecting that this structure is cuboctahedral, and not anticuboctahedral (twinned cuboctahedral) or hcp. The first is that the cuboctahedron allows a more symmetrical distribution of ligands than do the other two structures. Secondly, as will be pointed out in Sect. 3.1 below, it is possible to get an excellent fit to the MES spectra assuming only three different surface sites. With both these other close-packed structures, there is a much larger number of distinctly different surface sites, both with respect to the gold coordination as well as to the site symmetry. Although these arguments are only suggestive, further arguments will also appear in Sect. 3. Note also the EXAFS results on supported bare gold clusters [83, 84], where even for average cluster diameters as small as 1.1 nm, i.e. smaller than for Au55, only the fcc structure was observed.

2.3   Differential Scanning Calorimetry Results

DSC measurements on solid Au_{55} [57] (see Fig. 2) show a single sharp exotherm at 429 K, corresponding to the decomposition of Au_{55}, and a smaller endotherm at 474 K, which corresponds well with the known melting point of $Au(PPh_3)_2Cl$. Although many gold cluster compounds decompose at similar temperatures with loss of phosphine ligands [85, 86], this is further evidence

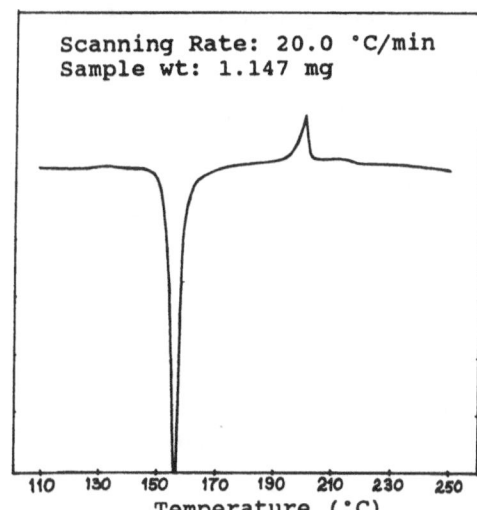

Fig. 2. Differential scanning calorimetry on solid Au_{55} between about 380 K and 530 K

that Au_{55} is a stoichiometric molecular compound. A mixture of several clusters, or a nonstoichiometric colloidal gold material, would not show a single sharp decomposition exotherm. The lack of further DSC peaks indicates that the decomposition observed here in the solid state follows the same route as that found by ^{31}P NMR to occur in solution [26, 16], into $Au(PPh_3)_2Cl$ and metallic gold. Thermogravimetric analysis has shown that a mass loss of about 12% takes place from Au_{55} as the temperature is increased after decomposition; this is consistent with the loss of PPh_3 ligands from the $Au(PPh_3)_2Cl$ formed, to give $Au(PPh_3)Cl$ [58].

From the area under the DSC exotherm, the two-center Au-Au binding energy per mole was calculated to be 76.1 kJ for Au_{55}, as compared to 61.3 kJ for bulk gold [57]. This is in agreement with the stronger binding implied by the reduced gold-gold distance found in Au_{55} by the EXAFS measurements. These results also fit a simple exponential relationship between bond length and bond strength in metal clusters [87], implying similar metal atom hybridization and degrees of metal atom core overlap [65] in Au_{55} and bulk gold. Other Au systems clearly different in hybridisation, such as the Au_2 dimer [88, 89] and long-range interactions between gold atoms with $5d^{10}$ electronic configurations [90], do not follow the same relationship [52]. As we will see in the following sections, there is considerable further evidence that the cluster core does indeed show metallic binding.

3 The Thermal Behavior of Au$_{55}$

3.1 The Thermal Behavior at Low Temperature

In systems containing a number of physically inequivalent sites, Mössbauer
effect spectroscopy (MES) can often allow the determination of the properties of
the individual sites. This also proved to be the case here [24].

Analysis of ^{197}Au MES measurements as a function of temperature between
1.25 K and 60 K on the Au$_{55}$ system, employing the transmission integral

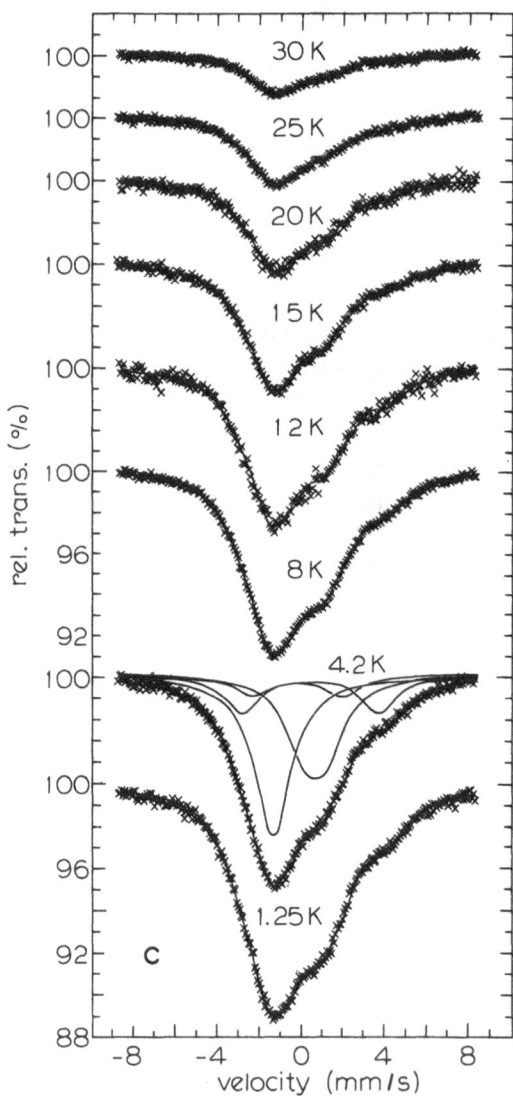

Fig. 3. Mössbauer spectra of Au$_{55}$ at
several temperatures between 1.25 K
and 30 K, with transmission integral fit
[24]. The subspectra at 4.2 K are loren-
ztian simulations to indicate the singlet
and the three doublets, and are only
meant as a guide to the eye.

method, gave very satisfactory results using one unsplit Mössbauer absorption line for the 13 core sites and three quadrupole split (Q.S.) doublets for the three distinct surface sites (Fig. 3). The fact that the experimental results could be fitted with a physical picture which allowed only 11 free parameters (4 I.S.s, 4 line intensities, and 3 Q.S.s) already constituted a compelling argument that the assumed structure was correct. These experimental spectra-are not consistent with the number of gold atom sites that would be present in an icosahedral or polyicosahedral cluster structure.

The four sites of Au_{55} exhibit different line intensities, and from the relative site occupations, the Mössbauer f-factors for the different sites could be calculated [24], using standard techniques [91]. These, in turn, could be related to effective Einstein (or Debye) temperatures θ_E (or θ_D) associated with the vibrations of the individual sites. An unexpected consequence was that the three surface sites could not be described by a single $\theta_E^{surface}$, meaning that the use of a θ_E^{bulk} and a $\theta_E^{surface}$ [92, 93, 94] was insufficient to explain these results. In addition, the f-factors for all four sites were considerably lower at 1.25 K than that for bulk gold at the same temperature.

A further result of this analysis, as shown in Fig. 4, is that while the relative spectral intensities are determined by the individual site f-factors, the temperature dependence of all the sub-spectra together in the temperature range studied is determined by the motion of the center of mass of the whole Au_{55} cluster. This can be seen by the uniform decrease of the total intensity with increasing temperature, without any visible change in the general shape of the spectrum. In effect, this means that the f-factors for the individual sites must be multiplied by an f-factor due to the motion of the whole particle [24]. See also Refs. [95, 96, 97], where this concept was originally developed. The use of such an inter-cluster f-factor, in addition to the usual intra-cluster f-factor, also resolved the problem of the apparent deficiency in the total f-factor at 1.25 K when compared to bulk gold.

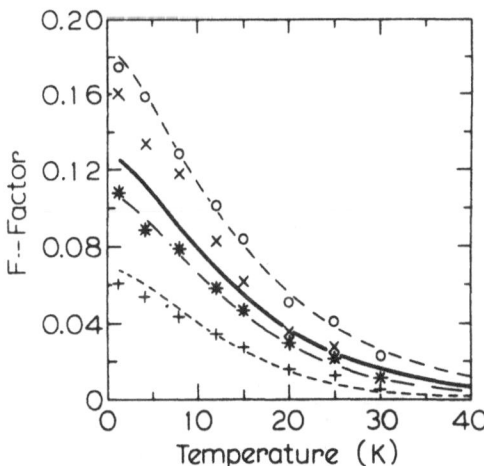

Fig. 4. The temperature dependence of the relative intensities of the MES sub-spectra [24, 25, 42]. The fit is discussed in the text

3.2   A Model Description of the Thermal Behavior
      at Low Temperature

Starting from the assumed cuboctahedral structure, utilizing a simple model of point masses and massless springs and completely ignoring the ligands on the surface, the eigen-frequencies and the degeneracies for the various gold sites have been calculated [25]. Ignoring the ligands is a valid first approximation, since it has been shown by FIR-spectroscopy [98] that the vibrational and rotational modes for Au_{55} lie well above 100 K. The original calculation was done with the interatomic distance of bulk gold and the spring constant determined from the bulk modulus of gold. The calculated Θ_E's were about 10% smaller than the measured values, averaging over all the sites. The only free parameter used in this calculation was the unknown "spring constant" for the motion of the center-of-mass of the cluster in the matrix of amorphously interlocked ligands, i.e. for the three inter-cluster degrees of freedom. This single free parameter was determined by fitting the temperature dependence of the f-factors, as shown by the drawn curves in Fig. 4.

A recent recalculation [43], utilizing the interatomic distance obtained from EXAFS [39, 40] and the spring constant corrected for the higher binding energy extracted from the DSC results [57] reduced the difference between the calculated and measured Θ_E's to about 2%.

The success of this simple four site model in predicting the relative site intensities using only one free parameter, taken together with the calculated linear dependence of the f-factors on the coordination number, is yet another confirmation of the assumed structure of Au_{55}.

As an additional point, the f-factors calculated for the four sites of Au_{55} [25], as well as those for several smaller and larger magic-number and non-magic number cuboctahedra, for most of which no real examples of gold cluster compounds exist, exhibit a linear dependence on the coordination number of each site. Several of these are shown in Fig. 5. Thus, the coordination numbers of

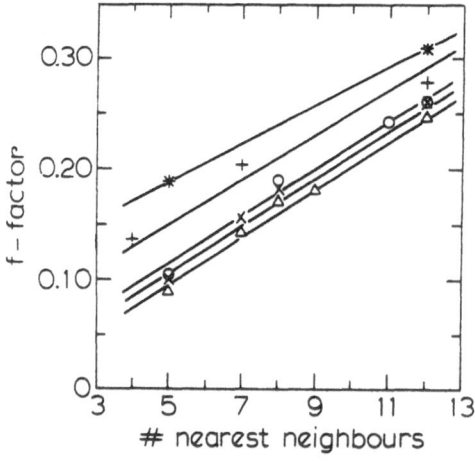

Fig. 5. The MES f-factors versus site coordination numbers calculated for cuboctahedral Au_{13} and Au_{55} as well as several larger magic-number and non-magic-number cuboctahedral cluster sizes. The symbols are defined as follows: $Au_{13} = *$, $Au_{19} = +$, $Au_{43} = \bigcirc$, $Au_{55} = \times$, and $Au_{147} = \triangle$

the individual gold sites in the cluster are physically meaningful, and play an important role in describing the cluster behaviour.

3.3 The Low Temperature Specific Heat of Au$_{55}$

The calculated specific heat for Au$_{55}$ at temperatures between 2 K and 60 K, based on this model, is shown in Fig. 6. Almost equally good agreement has been obtained for the specific heats and the f-factors for the cluster compounds Au$_9$L$_8$ and Au$_{11}$L$_7$X$_3$ (with X = SCN) [25].

Note that the full curve drawn in the figure corresponds to the specific heat calculated with *no free parameters*. While the agreement of this model to the specific heat is rather impressive, it is in fact the 3 inter-cluster degrees of freedom, i.e. motion of the whole cluster, which govern the phonon contribution to the specific heat at temperatures below about 15–20 K [99].

This in itself is proof of the importance of the center-of-mass motion in treating the thermal properties of small particles at low temperature.

An additional piece of information provided by this calculation is that the mean square displacement of the gold atoms on all the sites except for the PPh$_3$-coordinated sites are primarily, though not exclusively, radial. The mean square displacement of the gold atoms on the PPh$_3$-coordinated sites, on the other hand, have an in-plane (i.e. tangential) mean-square displacement which is not only larger than in the radial direction [94, 100], but also exhibits significant in-plane anisotropy. This may be related to an incompletely developed "soft mode" in the cluster [101], which might be considered as a precursor to a transition from the observed close-packed structure to the icosahedral structure, predicted theoretically by Sawada et al. [102].

The site f-factors are strongly sensitive to the details of the structure of the cluster. The fact that a model which completely ignores the type of ligand and

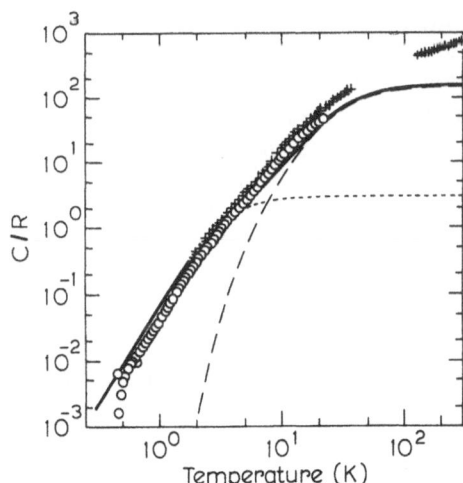

Fig. 6. The specific heat of Au$_{55}$ at low temperatures, with the specific heat—calculated with no free parameters—included as a *full line*; the *fine dashed line* is the inter-cluster contribution, the *coarse dashed line* the intra-cluster contribution

the bonding directions works so well in predicting these f-factors is a further strong indication that the Au–Au bonding in Au_{55} is non-directional and probably of metallic nature.

It should be pointed out here that the measured specific heat [25, 54, 55] of Au_{55} showed no trace of a linear term which would normally indicate the presence of an electronic contribution to the specific heat, at least down to 60 mK. We will return to this point in Sect. 4.5.

The behavior of the specific heat in external magnetic fields at temperatures below 300 mK [54] is quite puzzling. We will also return briefly to this point at the end of Sect. 4.5.

3.4   A Surface Mode Description in Relation to the MES Results

The experimental observation that one has different "Debye temperatures" for the three distinct surface sites of the Au_{55} cluster makes the use of a continuum-model picture for discussing the thermal behavior questionable. Indeed, for such small particle sizes, where the surface structure is so manifest, the use of the concept of "surface modes" becomes dubious, and is certainly inadequate to explain the observed temperature dependence of the f-factors. None the less, it has proven possible to describe the low temperature specific heat of Au_{55} quite well using such a continuum-model, when the center-of-mass motion is taken into account [99].

The average of the effective Einstein temperatures corresponding to the proper combinations and weightings of the three surface sites of Au_{55} do not agree with those measured by Kostelitz et al. [103] by LEED experiments at room temperature on flat gold surfaces. But the surface sites of a flat gold surface have both different (average) coordination and interatomic distances than is the case for Au_{55}. Related to this is the difference in surface tension between a small diameter "sphere" and a flat free surface. Furthermore, it should not be surprising that measurements done at 300 K differ from those carried out below 30 K. An EXAFS determination of the effective $\theta_D^{surface}$ of particles of average diameter down to 2.0 nm [94], also measured at room temperature using θ_D^{bulk} as reference, were also in disagreement with the MES results.

It is unfortunate that the low temperature EXAFS measurements of Marcus et al. [40] were interpreted without considering the inter-cluster motion, since an independent direct comparison to the MES results would have been useful.

3.5   The Melting Temperature of Au_{55}

Over the years there has been a lot of interest in the decrease in melting temperature of small metal particles, relative to the bulk, theoretically [104, 105, 106] as well as experimentally [107, 108, 109, 110]. Extrapolations from high temperature measurements have suggested that, for gold particles with a diameter of less than about 2 nm, the melting point would be 0 K [107, 105, 106, 109]. That is, gold particles of this size would always be in the form of "liquid droplets". A recent study [111] has questioned these results, claiming a melting temperature for Au_{13} of about 680 K. The situation concerning the melting temperature of small bare gold particles is, at best, unclear at the moment.

Sawada and Sugano [102] have also recently calculated possible structural fluctuations in bare Au_{55} clusters, and claim that for an uncharged Au_{55} cluster, the cuboctahedral structure at room temperature is energetically unstable toward the formation of the icosahedral structure. This is in direct conflict with the EXAFS results of Balerna et al. [83, 84]. For bare gold clusters in the average diameter range from 1.1 nm to 6.0 nm, deposited on a mylar substrate, only the fcc structure was observed, with the inter-gold distance being size dependent. We should point out here that the inter-gold distance that they obtained for clusters of about the size of Au_{55}, while smaller than for bulk gold, were larger than observed for Au_{55} [39, 40, 41]. We will return to this point in Sect. 4.3 below.

The stability of the Au_{55} cluster compound is of course supplied at least in part by the ligand shell, and this complicates a direct comparison of the above theoretical predictions for imaginary bare Au_{55}. But experimentally, the presence of an MES spectrum consisting of a superposition of distinct lines of the natural linewidth for gold from four different structural sites also constitutes proof that gold core of the Au_{55} cluster is, at least on the time scale of the MES measurements (i.e. 0.1 ps $< t <$ 10 μs) [112, 113], a *solid* up to temperatures of at least 30 K. Surface melting on this time scale can also be refuted for Au_{55} for the same reason. The same has also been observed by MES on the water soluble compound Au_{55}^* [46].

3.6 Sample Preparation

The MES measurements cited above were performed on samples prepared by placing the Au_{55} powder directly into a metallic absorber holder, sealed off with aluminium foil. The resultant spectra showed no trace of metallic gold mixed with the Au_{55} material, which would have been the case if there was serious ligand loss. Since the entire MES spectrum of Au_{55} had practically disappeared at 40 K, appreciable bulk gold contamination could not be present, since bulk gold, with a θ_D of nearly 185 K, would then have become clearly visible, if present [114, 115, 32].

As a check on the standard technique that is often used for other types of measurement [54, 33], a Mössbauer absorber was also prepared by dissolving Au_{55} in CH_2Cl_2 and allowing it to evaporate in the absorber dish. Subsequent MES measurements [45] were carried out at temperatures up to 60 K. A spectrum identical (within statistics) to that of Au_{55} plus a singlet corresponding to about 5% metallic gold was obtained at 4.2 K. The fact that a measurement done at 40 K didn't show an apparent increase of the relative percentage of metallic gold means that the metallic gold present is also in the form of particles small enough to also be subject to a large center-of-mass motion [114, 115, 32]. This lack of large-scale coalescence could be due to the very rapid evaporation of the absorber sample, which might leave the bare gold particles isolated in the matrix of Au_{55}. None the less, we can be certain that preparing samples by dissolving Au_{55} and then evaporating it on a substrate can lead to the damaging of at least part of the material. Due to the large amount of undamaged Au_{55} present in the sample, it is impossible to extract information concerning the other decomposition products. On the other hand, these results would indicate that sample preparation by the dissolution/evaporation route for Au_{55} [54, 33] may lead to considerable sample deterioration, a point already emphasized by Fauth et al. [59]. This may account for the discrepancy between the plasma desorption mass spectroscopic results [33] and most of the other experimental evidence on Au_{55}; another explanation could be sample damage caused by the extremely high energies involved in cluster vaporization by impact with ^{252}Cf nuclei.

4 The Electronic Structure of Au_{55}

4.1 Electrons in a Small Particle of Gold

Before presenting experimental results concerning the electronic structure of the Au_{55} cluster compound, it is necessary to first examine several concepts relating to the electronic structure of small gold particles in general, and to try to see how these differ from the bulk.

Bulk gold has the fcc structure and behaves much as a normal monovalent metal. The calculated Fermi energy is about 5.51 eV [116]. The electronic structure can be represented as $5d^{10-x}6s^{1+x}$, with $x \ll 1$, i.e. the $5d$ band is slightly less than full, and the $6s$ band slightly more that half filled, the very broad $6s$-band overlapping and hybridizing with the rather narrow $5d$-band [65]. In fact, this is an oversimplification, since there will also be p-character present as well, which will also play a role in the hybridization process. Despite the strong hybridization typical for transition metals, the electronic behaviour can be fairly well described on the basis of nearly free electrons: there is plenty of room in the valence band and the effective mass of the electrons in bulk gold is very nearly unity.

One can expect a number of changes in the system when the size of the sample becomes small. Below we will see that the gold atoms in the core of Au_{55} reflect many characteristics of the metallic state. For this reason we will look specifically at Au_{55} as an example of a "small metallic system", and will later examine the validity of this approach.

Normally one might consider only the $6s$ electrons as being delocalized into a broad s-like conduction band. A general observation makes it essential, in the case of gold atoms, to consider as well the 10 $5d$ electrons. Experimentally, the variation with cluster size of electronic properties can be explored by XPS measurements. These properties can be divided into those which converge rapidly toward bulk-like values ($Au4f$ binding energies (b.e.) and $Au5d$ valence band (VB) splitting [74, 75]) and those which change much more slowly (such as $Au5d$ VB shape as a function of photon energy [117]). Theoretically it has been suggested [117] that clusters of only 30 gold atoms already develop bulk character, while nuclearities of about 250 are necessary for properties which are dependent on long range crystal structure (i.e. on the crystal momentum vector k becoming a good quantum number). XPS measurements on supported bare gold clusters [117] show the development of an apparently well defined $5d$ band for clusters of only 10 to 15 atoms and approach bulk character already for clusters of 50 to 100 atoms. This means that we must consider 11 electrons per atom as being somewhat delocalized into overlapping d and s valence bands, i.e. 605 per Au_{55} cluster.

We could try do a simple "particle in a cubic box" quantization for the $6s$ and $5d$ electrons, giving degeneracies of 2, 12, 24, 16, etc., or alternatively we could quantize in a spherical well, giving degeneracies of 2, 8, 16, 32, etc. Since the "diameter" of Au_{55} varies by nearly 25% as a function of direction, a "spherical well" might be less than satisfactory, but then again, our "cubic box" has had its 8 corners truncated. In both cases we expect a relatively small number of highly degenerate levels. But the degenerate electronic levels must eventually approach the corresponding bulk bands, and the limited size of the cluster and the accompanying boundary between occupied and unoccupied "lattice sites", i.e. for the gold atoms of Au_{55} in their fcc lattice, will lead to at least partial lifting of degeneracies.

The degeneracies can also be affected by Coulomb interactions and quantum tunneling between neighboring clusters, by the presence of the ligands on the surface of the cluster, and by correlation effects within the cluster. Finally, we must not forget to reinsert the hybridization between $5d$, $6s$, and $6p$ electrons which we neglected in order to simplify the discussion. In the end, without an actual calculation of the level structure, which is still at present outside the limits of possibility, we can do little more than try to estimate the relevant splittings on the basis of the experimental results. At this point we can only say that although the splitting between the bands may be very large, of the order of thousands of Kelvins, the splittings between the sub-levels within a band will be much smaller, probably of the order of a few Kelvins to a few tens of Kelvins.

We can now ask ourselves what the effect of the ligands is on the number of valence electrons in the system. IR Au–Cl stretching frequencies [16] and XPS (Cl2p) binding energies (b.e.s) [48] indicate that the Au–Cl bonds in Au_{55} are intermediate in nature between those of the covalent compound $Au(PPh_3)Cl$ and the partly ionic compound $Au(PPh_3)_2Cl$ [118]. We can thus expect that the six chlorine ligands will remove between three and six electrons out of the core.

Based on a comparison of XPS binding energies (b.e.s) of the Cl- and PPh$_3$-coordinated surface atoms of Au$_{11}$L$_7$X$_3$ (with X = I and Cl) [76] and Au$_{55}$ [48], we can estimate that the twelve PPh$_3$-ligands of Au$_{55}$ will remove another three to six.[1] This means that less than twelve valence electrons will have been removed from the cluster core, constituting a maximum 2% change in the total number of delocalized electrons in the Au$_{55}$ cluster.

If the energy levels of these electrons were *uniformly* distributed in energy up to the Fermi level of 5.51 eV, they would have an average energy separation of the order of 9.1 meV (106 K). Experimentally, however, this is not the case. From XPS measurements (Sect. 5), the Au$_{55}$ clusters have valence bands showing a strong similarity to the bulk (i.e. with a much smaller level separation than 106 K), with the Fermi level located at the upper edge of the valence band.

Following this analysis, we will use the term *metallic binding* to mean the behaviour due to the delocalization of the valence electrons into bands of energy levels. We will use the term *metallic character* in the context of the transition from the mesoscopic regime, with discrete but closely spaced energy levels, to the continuum bands of the bulk. When the relevant discrete energy level splittings become smaller than k$_B$T, for example, we will consider that metallic character is present for the *particular property* being considered.

Just as the energy splitting within a band of levels will effect the physical properties of the cluster, so will the density of the sub-levels, the filling of the individual sub-levels, and the filling of the band of levels. For possible collective motion of the electrons within the cluster, the band must not be completely filled. In the case of Au$_{55}$ this point is not completely clear.

The Fermi level is basically determined by the total electron density in the system. Although defining a Fermi level cannot be done in a straight-forward way unless $n \rightarrow \infty$, it should be roughly the same for the cluster as for the bulk, i.e. about 5.5 eV. In the case of the cluster in the mesoscopic region, it is perhaps better to speak of the highest filled level than of the Fermi level. The distance between the highest filled (valence or HOMO) level and lowest unfilled (conduction or LUMO) level will determine the temperature range in which collective intra-cluster electronic behaviour is possible. In this sense it will also determine which properties belonging to collective behaviour have become bulk-like within an individual cluster.

When small clusters are placed adjacent to one another, but at random with respect to inter-cluster distances and orientations, as is the case in the amorphous arrangement of the clusters of Au$_{55}$, the interactions between the individual clusters will lead to small shifts of the intra-cluster energy levels which are slightly different for each cluster. Since both macroscopic as well as microscopic

---

[1]This highlights contrasting views of the PPh$_3$ ligand, which acts as a σ-donor and a π-acceptor. Extensive IR studies of substituted metal carbonyl compounds have shown that the π-acceptor ability of PPh$_3$ is rather weak [119]. However, the clear interpretation of XPS binding energy measurements on gold phosphine complexes is that the overall electronic balance between the two effects is a net electron withdrawal from P-ligated Au atoms [76]. See also Sect. 5.2.

measurement techniques measure an average over the clusters present in the system, the intra-cluster energy levels, even if relatively sharp, may appear to be spread into bands, i.e. may appear somewhat as a continuum. This point will be of importance in the discussion of the electrical conductivity in Sect. 4.5 below.

In recent publications [120, 121, 122, 123] it has been shown that both the ionization potentials and the optical properties of bare and uncharged mercury clusters in a molecular beam experiment demonstrate a gradual size dependent evolution of metallic properties, starting at about 13 atoms and already bulk-like at about 70 atoms. It has been *predicted theoretically* [124] that plasmons should begin to develop for such mercury clusters at about Hg_{15}. We should keep this in mind in the discussion of the electronic properties of Au_{55}.

4.2 The Metallic Binding of the Cluster Core

For reference, the MES I.S. value of bulk metallic gold, relative to Au in Pt at 4.2 K, is shown in Fig. 7 by the filled circle [125]. Gold molecular compounds which have such negative I.S. values are strongly ionic and often quite unstable as well.

In the same figure, the Q.S. and I.S. values of Ph_3PAuCl and $(PPh_3)_2AuCl$ (as reference material) are given as a vertical cross and a diagonal cross, respectively. The PPh_3-coordinated sites of $Au_{11}L_7(SCN)_3$ and the Au_{55} cluster are shown by the open and centered triangles, respectively. The Cl-coordinated sites are shown by open and centered squares. The central site of $Au_{11}L_7(SCN)_3$ is given by the open circle and the core sites of Au_{55} by the centered circle. The centered diamond is used for the bare surface sites. It should

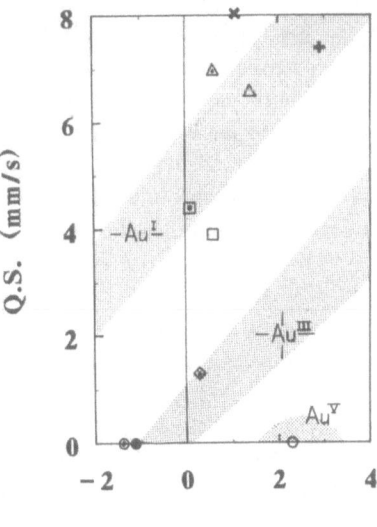

Fig. 7. Values of Q.S. versus I.S. for the three sites of $Au_{11}L_7(SCN)_3$ (*open figures*), for the four sites of Au_{55} (*centred figures*), and for bulk gold (*filled circle*), shown against a background which gives the approximate ranges of Q.S. versus I.S. values for most gold compounds. The *two crosses* are for reference compounds explained in the text

be noted that the I.S. value of the core sites of Au_{55} lies quite close to that of bulk gold, in strong contrast to that of the central site of $Au_{11}L_7(SCN)_3$, but this is *not* due to contamination with bulk gold, as was pointed out in Sect. 3.6 above. The I.S. value of the core sites is an indication that they have an electronic 6s density similar to, but slightly less than, bulk metallic gold. Furthermore, the various surface sites of Au_{55} all have significantly smaller values of the I.S. than the $Au_{11}L_7(SCN)_3$ cluster compound. Similar results have been found for the methylated compound Au_{55}-tolyl and for all but the Cl-coordinated site of the water soluble compound Au_{55}^* [46].

In a comparison of the I.S. of bulk gold with the I.S.s of the central sites of the smaller centered gold clusters [1, 2, 3, 4, 24, 25], the latter are all greater than $+2.0$ mm/s, while bulk gold has an I.S. of -1.22 mm/s. These very large positive values for the I.S.s have been attributed [1, 25] to a transfer of charge from the central site to the peripheral ligands, the central atom then having an effective valence of $+5$. In sharp contrast to this is the I.S. of -1.4 ± 0.1 mm/s of the central 13 core atoms of Au_{55}, i.e. very nearly the same as that of bulk gold.

4.3   The Isomer Shift of the 13 Core Sites of Au_{55}

This particular value of the I.S. presents a problem. The I.S. observed for the 13 core atoms, which have the same coordination as bulk gold [24], is more negative than the I.S. of bulk gold metal itself. It should be noted that for ^{197}Au, the second order Doppler shift (S.O.D.) is about 2 orders of magnitude smaller than our estimated limits of error. If we look at the possible contributions of the 6s, 6p, and 5d electrons to the effective charge at the gold nucleus, as registered by the I.S. [125], it turns out that the 6s electrons are responsible for the major part, amounting to about $+8.0$ mm/s per 6s electron. The screening effect of the 6p electrons on the s electrons is almost completely cancelled by the contribution due to the relativistic density of the 6p electrons within the nuclear volume, and can be safely neglected. The 5d valence electrons have a net screening effect on the s electrons which is roughly equal to -1.6 mm/s per 5d electron. A change in the character of the electrons, from s-like to d-like, would also have the effect of reducing the I.S.

The decrease in Au–Au distance in Au_{55} relative to bulk gold, obtained from the EXAFS [39, 40, 41] measurements, would be expected to compress the 6s electrons into a volume about 10% smaller than in the bulk, which would lead to a change in the I.S. of about 0.7 mm/s in the positive direction relative to the bulk [126]. We should thus not expect a change of 0.2 mm/s in the negative direction, as observed experimentally for the core sites [24].

In order to understand this effective loss of 6s electron density by the core site atoms, use has been made [44] of the model introduced by Citrin and Wertheim [65] to explain the negative surface atom core level shifts, relative to the bulk, observed in the XPS spectra of metals surfaces [127]. Their model used gold as an example to demonstrate that, due to the lower coordination of the surface sites, the wave functions of these sites become more localized, i.e. more d-like, necessitating a small net flow of electrons from the bulk metal to the surface layer.

The experimental results obtained for low nuclearity bare gold clusters on non-conducting substrates show that the individual atom core b.e.s are slightly larger than in the bulk [74, 75]. This is an unexpected result, since one can think of the cluster as an intermediate species between the surface and the bulk. In a cluster as large as Au_{55} we can distinguish between the surface and core atoms, and in the absence of theoretical calculations we may tentatively assume that the surface versus core b.e. shift for a bulk metal will be analogous to the corresponding shift in a cluster, as a first order approximation. See Sects. 5.1 and 5.2 as well.

In our clusters, the 13 core atoms constitute the "bulk" metal, relative to the 42 surface atoms. Furthermore, the average surface coordination number of 6.57 for the cluster is lower than the average of 8.5 for the most stable surface faces in a flat gold surface (i.e. $1/2(100) + 1/2(111)$) [127]. Thus, extending the Citrin and Wertheim model to Au_{55}, much more charge would need to leave each individual gold atom in the Au_{55} core than in the usual situation with the surface of a normal macroscopic sample. It is this effect which can provide a qualitative explanation for the huge drop of $6s$ electron density at the core sites.

On the other hand, a comparison of the inter-gold distances obtained by EXAFS on the Au_{55} clusters [39, 40] with those obtained on bare gold clusters from EXAFS [83, 84] and electron diffraction [105] shows that the reduction of inter-gold distance in Au_{55} is twice as large as in the bare clusters of the same size, relative to the bulk. Using the argument of Citrin et al. again [65], the loss of about 12 valence electrons to the ligand shell of the Au_{55} cluster might well explain the difference in inter-gold distance between the bare and ligated clusters as being due to increased (core level) binding as a consequence of the decrease in Coulomb repulsion. We will also return to this point in Sect. 5.2.

4.4 The Metallic Binding of the Surface Sites

Let us maintain the assumption that the 13 gold atoms in the core of the Au_{55} cluster have metallic binding, i.e. delocalized valence electrons. The further assumption that the bare surface sites, since they are in direct contact only with other gold atoms, also have metallic binding follows almost automatically. Although less obvious, if the ligated gold surface atoms are also assumed to have delocalized valence electrons, one might then apply the model of Miedema and v.d. Woude [66, 128] to estimate the I.S. shifts of all three surface sites [67].

This simple empirical model for predicting the I.S. shift of a Mössbauer nucleus placed in a metallic system (alloys, as well as intermetallic compounds), uses differences in the tabulated macroscopic work functions and bulk moduli to model differences in the microscopic electronegativities and electron densities at the areas of contact of unlike atoms at the Wigner-Seitz cell boundaries [129]. These in turn are related to the changes in I.S. due to charge transfer and to the rehybridization of the electrons at the cell boundaries via geometrical factors describing the contact areas, and using empirical constants determined for each Mössbauer isotope individually.

To simplify the application of this model to our cluster compound, we will completely ignore the presence of the ligands, since they might be expected to lead only to minor local corrections to the I.S. values [24, 96]. Furthermore, under the assumption that all the surface atoms are metallic, it is possible to make a drastic simplification to the model of Miedema and v.d. Woude. The "unlike atoms" in contact with the surface sites are vacuum, i.e. missing neighbors, and as such will not be available to transfer charge to or from the surface sites. The rehybridization, on the other hand, *will* take place, as we have already mentioned above [65], and the corresponding geometrical factor is fully determined by the number of nearest gold neighbors. This leads, with *no adjustable parameters*, to the values given in Table 1. These predictions for the I.S. values of the three surface sites are remarkably good, and in the correct order of relative magnitude. The core site I.S. value was used as the reference level, and a standard correction of -0.8 mm/s, normally used to bring model values into agreement with experimental values, has been included there as well.

Despite controversies about the physical basis of the model [130, 131, 132], it has proven to have strong predictive value [133, 134]. The surprisingly good agreement between experiment and this extreme simplification of an empirical model constitutes an additional strong argument that the *entire gold core* of Au_{55} does indeed exhibit metallic binding. That the ligands also seem to be of only secondary importance here in determining the value of the I.S., as was postulated, is a point which merits further investigation [46].

4.5   Other Experimental Evidence for Metallic Binding

The results of temperature and frequency dependent measurements of the DC and AC conductivity reported by van Staveren [29, 61, 62, 63, 30] could be successfully analysed in terms of a modification of the standard models [135, 136, 137, 138] for granular metals exhibiting variable range hopping [50]. These models are based on the idea of an array of *identical conducting spheres* embedded at random distances in a non-conducting matrix. The amorphous arrangement of the Au_{55} clusters insulated from one another by their non-conducting ligands fits this model quite well. Although in the standard models for variable range hopping one considers hopping between nearest and next nearest neighbors, as utilized by van Staveren [50] only nearest neighbor hopping was considered, i.e. "fixed range hopping". Both Au_{55}^* and Au_{55}-tolyl gave similar results. This has recently been described by Brom et al. in terms of a Mott-Hubbard hamiltonian [63].

We can consider this as a further confirmation of the results which have already been extracted from the microscopic measuring techniques mentioned above, supporting the idea of the metallic binding in the Au_{55} cluster. Were the highest occupied band of intra-cluster energy levels completely filled, it would be difficult to imagine electrical transport by electron hopping from an effectively insulating cluster to the next.

On the other hand, the fact that at temperatures below 77 K the impedance had become too large to allow measurement of the electrical conductivity is a strong indication that correlated electronic motion is no longer taking place, or at least is strongly hindered. Taking a temperature splitting which would give a Boltzmann factor of 0.1 as a rough estimate of the interband splitting within the cluster, one obtains a value of about 35 K (~ 3 meV).

A difference of two orders of magnitude has been measured in the DC conductivity of Au_{55} and Au_{55}^{*} at 100 K [29, 30]. From this, and the MES observation [46] that the total electron count in the metal core of Au_{55}^{*} may be about 2 less than in Au_{55}, it would appear that there is a very nonuniform distribution of levels, with a much larger energy splitting for the HOMO level of Au_{55}^{*} than for Au_{55}. Furthermore, the density of states in these levels must be very low.

By studying the Au_{55} system by means of ^{31}P NMR, one would hope to be able to obtain additional microscopic information about the electronic behavior. Unfortunately, NMR experiments proved to be non-trivial [29], since the ^{31}P resonance was extremely weak. This has been taken as an indication that metallic shielding may still be incomplete.

The main result extracted from ^{31}P NMR was that the Au_{55} clusters do not exhibit a normal Korringa relation. Rather, there is an indication of the sort of general two-level behavior often seen in disordered glassy systems. This does not appear to be in disagreement with the results reported above, especially when one considers the modification of the intra-cluster energy levels due to the inter-cluster interactions.

Magnetic susceptibility and magnetization measurements have been carried out [64, 53]. The Au_{55} clusters have a diamagnetic susceptibility which is more or less constant down to around 77 K, gradually lose their diamagnetism, and below about 30 K appear to become paramagnetic, with a measured magnetic moment equivalent to about 1/3 unpaired electron/cluster. The magnetization measurements are field independent at 300 K, 77 K, and 20 K for fields between 0.1 T and 0.65 T. In pulsed fields up to 40 T at 4.2 K, the inverse susceptibility remained unsaturated.

As pointed out by van Ruitenbeek et al. [139, 140], the orbital contribution to the magnetic behaviour in systems of finite size can in general be expected to be quite small. Therefore any appreciable change in the magnetic behaviour as a function of temperature must be due to the electron spins. In the Au_{55} system, the gradual change from diamagnetism above about 70 K to paramagnetism below about 30 K, with a clear Schottky-like anomaly around 10 K, may be due to gradual localization of the last delocalized electrons into the ground state configuration, i.e. this is a further indication that the splitting between the highest occupied and the lowest unoccupied energy state must be of the order of a few tens of Kelvins. This is in accordance with the previous results.

It is important to note that in principle Au_{55} is an odd electron compound. Thus, we would expect Curie or Curie-Weiss paramagnetism at low temperatures for the unpaired electron spin. At higher temperatures, the large diamagnetic contribution from the many other electrons in the system dominates. On the other hand, a value of the magnetic moment of about 1/3 unpaired electron per cluster has been observed. This could be due to an approximately uneven number of electrons being taken up by the ligands, leaving a nearly even electron count in the cluster core. It could also be indicative of cluster damage and/or impurities in the sample. Remeasurement on freshly prepared and well filtered samples is in progress [141].

If metallic conductivity within the cluster should be present, with mobile, delocalized electrons, one might hope to see a linear electronic term in the specific heat [54, 56]. Considering the results above, such a linear term should not be expected for Au_{55}.

The T^3 phonon term [25], corresponding to the center-of-mass motion of the clusters, can be safely extrapolated for temperatures below 2 K, and the linear term can be estimated from the known value of γ for bulk gold [142] in the same temperature range. These two terms would be approximately equal for a temperature of about 40 mK.

A linear term *has* been observed in the specific heats of the larger metal cluster compounds measured down to 20 mK [49, 56]. The value of the linear term of these cluster compounds, which have metal cores of Pt_{309} and Pd_{561}, is only a fraction ($\approx 1/3$) of the bulk value. We might extrapolate to Au_{55}, and use a fraction smaller than 1/3. But even using 1/3, the linear term would only become equal to the cubic term at about 15 mK.

The specific heat of Au_{55} has been recently measured between 60 mK and 3 K, as a function of external magnetic field [54]. The increase in the specific heat at the lowest temperatures was attributed to a possible Schottky tail from ^{197}Au nuclear quadrupole splitting.

The measured Q.S. values for the surface sites of Au_{55} give nuclear splittings of 21 mK, 13 mK, and 4 mK for the PPh_3 coordinated sites, the Cl coordinated sites, and the bare surface sites respectively. From these values, and the known site occupations, the nuclear quadrupole contribution to the zero field specific heat by these three two-level systems [143] has been calculated directly [144]. This value is 5 times as large as that experimentally observed [54]. The maximum value of a linear term in the specific heat of Au_{55} has been estimate to be no more than one fifth of the bulk value [144].

The observation of a linear term, if present, would thus appear to be practically impossible. None the less, we agree fully with the suggestion of Goll et al. [54] that there probably is no linear term for Au_{55}.

The low zero field C_v value which they observed is probably due to an exceptionally long nuclear quadrupole to lattice relaxation time [144], in conjunction with the relatively short times involved in the heat pulse method which they used [54].

4.6   ESR Measurements on Au_{55}

As mentioned above, Au_{55} is an odd-electron compound, and should therefore be ESR-active. However, we have been unable to observe reproducible ESR spectra from solid samples of Au_{55} [17, 52, 53]. A signal observed at room temperature in air gave a g-value of 2.00, linewidth 0.15 mT, but it disappeared on evacuating the sample tube for low-temperature studies, and no ESR signal could then be detected at temperatures down to 4 K. Although it is possible that the ESR spectrum was genuine, and that the sample desorbed ligands under high vacuum [77], it is much more likely that the signal arose from a volatile radical contaminant.

Recently, an ESR spectrum from Au_{55} at 4.4 K, with g = 1.9204 and linewidth 67 mT, has been reported [54]. The intensity of the signal followed the Curie-Weiss law, and the g-value and linewidth were independent of temperature. If the linewidth were determined by the electronic spin-spin relaxation time T_2, it would be expected to increase with temperature, so its temperature-independence suggests inhomogeneous broadening by unresolved hyperfine coupling to ^{31}P and ^{197}Au nuclei. Similar behaviour has been found in rhodium carbonyl clusters [18].

We can consider what is known about conduction electron spin resonance in gold particles. In a 5d metal with strong spin-orbit coupling, a large g-shift from the free-spin value of 2.0023 may be anticipated. The electronic relaxation time T_2 is expected to be very short, giving an ESR line too broad to be observed except at cryogenic temperatures. However, if gold behaves as an s-band metal (as evidenced by the plasma resonance absorption in the optical spectrum of gold colloids), a smaller g-shift would result. A theoretical g-value of 2.021 has been calculated [145, 146], but since this calculation gives an incorrect Fermi energy of 7.8 eV, its reliability is uncertain.

The experimental picture is also unclear. Dupree et al. [147] reported that gold colloids in the range 2 nm–4 nm (somewhat larger than our Au_{55} clusters) give ESR spectra with g = 2.26, linewidth 20 mT, between 180 K and room temperature. Monot et al. [148] subsequently found colloids of the same size range to give g = 2.0024 with a linewidth of only 0.6–0.9 mT at room temperature. A g-value of 2.11 has been obtained for Au foil below 20 K [149]. The whole subject clearly merits further experimental investigation.

Apart from relaxation time considerations, we can propose two other possible reasons why ESR might not be observed in solid Au_{55}. Firstly, the influence of local magnetic fields from the unpaired electrons on neighbouring clusters in the solid might broaden the spectra beyond detection. Secondly, there may be a disproportionation of charge in the solid to give even-electron species;

this would correlate with the electron-hopping found from the AC and DC electrical conductivity measurements. Further ESR experiments are in progress.

4.7 X-Ray and UV-Visible Absorption on Au_{55}

A comparison of the X-ray absorption edges of bulk gold and Au_{55} [39, 40] indicates that all the main features of the edge of the bulk metal are also exhibited by the cluster, the "white line" intensity showing qualitatively that the gold atoms in the cluster have a low mean oxidation state [150], closely approaching that in the bulk.

Marcus et al. [40] argue that this, taken together with a very weak bump around 520 nm in the optical absorption of Au_{55} in a tetrahydrofuran solution, when compared to similar measurements on colloids, is indicative of collective metallic behavior. 520 nm is the well characterized absorption wavelength for the plasma resonance in colloidal gold for particles of average diameters down to 1 nm [69, 59], showing delocalized electrons with correlated behaviour.

These results are also in accord with earlier UV-visible absorption results [57, 59] on Au_{55} dissolved in dichloromethane, as shown in Fig. 8. The bump seen in the cluster spectra is always very weak.

In more recent optical extinction measurements on specially filtered samples [60], the weak bump had disappeared. This bump may thus have been due to damaged clusters (see Sect. 3.6) or cluster aggregates which form colloidal inclusions in the sample. The main feature of the UV-visible spectrum of Au_{55} in solution is then a broad absorption extending across the whole visible region,

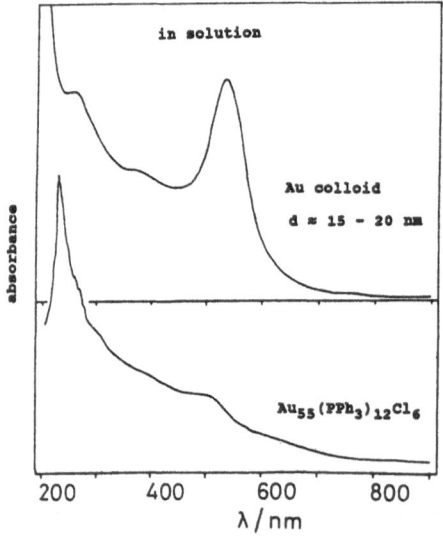

Fig. 8. Typical optical absorption spectra of Au_{55} and gold colloids in solution

from about 250 nm to at least 1350 nm. No narrow molecular-like absorption structures are observed in the range to 4000 nm [60]. The spectrum from Au_{55}^* in aqueous solution [60] is almost identical.

This behavior differs completely from the discrete one-electron absorptions of low-nuclearity metal cluster molecules [17]. Instead, it resembles the $5d \rightarrow 6s,6p$ interband transition of colloidal gold. This demonstrates clearly that the Au_{55} cluster has electronic energy levels which are closely spaced in a developing band structure, quite similar to colloidal gold. On the other hand, these electrons do not seem to show a collective behavior which would give rise to the plasma resonance.

Kreibig et al. [68] have recently proposed three possible explanations for the lack of the plasma resonance in Au_{55}. The first is that collective behavior doesn't occur, due to localization of the $6s$ electrons. This doesn't appear to be in agreement with the bulk of the experimental indications mentioned above, nor with that which will appear below in Sect. 5.4.

The second and perhaps most probable explanation is damping and broadening of the resonance, due to size dependent, single electron $5d \rightarrow 6p,6s$ interband transitions. Their explanation is that the discrete level structure of the Au_{55} cluster acts as an effective decay channel. In reducing the plasmon lifetime, it would also strongly increase the bandwidth of the resonance, washing out the resonance peak.

The third model is a shift to higher frequencies and concomitant broadening of the resonance peak, due to increased $6s$ electron density in the cluster. Here use was made of the expected increase of $6s$ electron density due to the decrease in cluster volume, as obtained from EXAFS [39, 40, 41]. But as we have pointed out above, despite the decrease in cluster volume, the I.S. of the core sites indicates a *decrease* of $6s$ electron density. The excellent prediction of the I.S. values of the surface sites, given above, utilizing a sizeable increase in d-character of the electrons associated with (in the vicinity of) the surface sites, means that the postulated blue shift and flattening should really be a red shift and concomitant sharpening of the resonance, which should make the resonance more visible, if present. The MES I.S. results thus refute this as a possible explanation.

As a further possibility, we can mention that the quantitative intensity C_{abs} of a plasma resonance absorption depends on the third power of the particle diameter a [151, 72]:

$$C_{abs} = \frac{8\pi^2 a^3}{\lambda} \Im\left(\frac{\varepsilon - 1}{\varepsilon + 2}\right)$$

where λ is the wavelength in the medium, and ε is the complex relative permittivity of the metal relative to that of the surrounding medium.

The plasma resonance absorption in a cluster as small as Au_{55} might simply be too weak to detect above the background interband absorption, which is quite intense at 520 nm, without the need to invoke any special broadening or damping mechanisms.

We have already referred, in Sect. 4.1, to the development of a plasma resonance in mercury clusters [124].

5   Photoelectron Spectroscopy on Au$_{55}$

5.1   XPS Results on Au$_{55}$

Photoemission is a powerful tool for the study of small metal particles and clusters, with a very rich literature on the subject extending over more than the last decade [74, 75, 76, 117, 152]. Both the interatomic binding as well as the shape and filling of the conduction band near the Fermi edge have been studied as a function of average particle size and of photon energy. In discussing XPS results it is necessary to realize that this is a final-state technique, which registers the electronic state of the system after the loss of the photoelectron.

In this section, we present XPS results [48, 49] which are highly relevant to the physics of this cluster system. These results and their interpretation are in sharp disagreement with photoemission results on Au$_{55}$ reported recently [47]. Both our results and those of Quinten et al. will be discussed in connection with the XPS data reported on smaller gold cluster compounds [76, 153] and on bare gold clusters supported on poorly conducting substrates [74, 75].

The XPS spectra were run on a VG-ESCA 3 photoelectron spectrometer, employing an Al K$_\alpha$ source (hν = 1486.6 eV). The XPS samples were made by pressing Au$_{55}$ powder onto a graphite substrate, which contributed only a very slowly increasing background in the valence band region. The samples were found to be unstable under X-rays for prolonged irradiation times. All spectra reported were taken within a few minutes (typically 5 to 10) from introduction into the analyzer chamber from an attached chamber in vacuo.

XPS data on Au$_{55}$ are collected in Table 2, together with literature results for smaller gold molecular clusters, and for bulk gold. Our Au4f peaks and the valence band spectrum for Au$_{55}$ are shown in Figs 9 and 10, respectively.

There are three main points which characterize the present XPS results and which will be discussed below. The first is that the four different Au sites in the cuboctahedral Au$_{55}$ structure have been resolved, each with its own characteristic binding energy. The second is that the 4f binding energies in Au$_{55}$ form a consistent series with the values for the smaller clusters. As was seen with the I.S. above, the central atoms in Au$_{55}$ are not so positively charged as in Au$_{11}$. And the third point is that the 5d spin-orbit splitting in Au$_{55}$ is larger than in the smaller Au clusters, and is quite close to the value for bulk gold.

A first difference between our XPS results and those of Quinten et al. is the lower statistics that our XPS spectra exhibit in comparison to theirs. The reason for the low statistics in our 5d peak lies in the relatively fast spectra acquisition deliberately used in order to prevent changes in the spectrum due to irradiation.

Table 2. A collection of relevant XPS parameters of Au_{55}, of several smaller gold molecular clusters, and of bulk gold [48, 76, 153]. The results quoted by Quinten et al. [47] for Au_{55}, as well as the results of Citrin et al. [127], have been included for convenience. a) our experimental ratios for Au_{55} are those fixed in the fit (see text); b) the trough depth is as defined in Ref. [153]

cluster type	ligand type	Au $4f_{7/2}$ (eV)	relative 4f ratios (exp)a	relative 4f ratios (th)	trough depthb	atomic ratios Au/P	Au/Cl	d-band split. (eV)	d-band shift	Refs.
LAuCl	L = PPh$_3$	85.7	—	—	0.23	1.01	0.84	—	—	[48]
L$_2$AuCl	L = PPh$_3$	85.15	—	—	0.22	0.53	0.82	—	—	[48]
Au$_9$L$_8$N$_3$	L = PPh$_3$ N = NO$_3$	85.2	—	—	0.22	1.14	—	1.8	—	[153]
	N = pic$_3$	85.2	—	—	0.23	1.14	—	1.9	—	[153]
	N = PF$_6$	85.2	—	—	0.23	—	—	1.9	—	[153]
Au$_{11}$L$_7$X$_3$	L = PPh$_3$ X = I	84.7	—	—	0.27	1.42	—	1.9	—	[153]
Au$_{11}$L$_7$X$_3$	center	84.53	—	—	—	—	—	2.1	0.52	[76]
	L = PPh$_3$	85.01	—	—	—	—	—	—	—	
	X = I	85.59	—	—	—	—	—	—	—	
	center	84.59	—	—	—	—	—	2.1	0.53	[76]
	L = PPh$_3$	85.07	—	—	—	—	—	—	—	
	X = Cl	85.77	—	—	—	—	—	—	—	
Au$_{55}$L$_{12}$X$_6$	L = PPh$_3$ X = Cl				0.295			2.4	0.3	[48]
	bare	84.0	24	24	—	—	—	—	—	
	core	84.4	13	13	—	2.8	4.3	—	—	
	PPh$_3$	85.3	12	12	—	P/Cl = 1.5	—	—	—	
	Cl	86.3	6	6	—	—	—	—	—	
Au$_{55}$L$_{12}$X$_6$	core	84.63	20.35	13	0.11	—	—	2.4	0.48	[47]
	surf.	84.25	34.65	42	—	—	—	—	—	
Au$_{bulk}$ (flat)	bulk	84.0	—	—	—	—	—	2.4	0.0	[127]
	surf.	83.6	—	—	—	—	—	—	—	

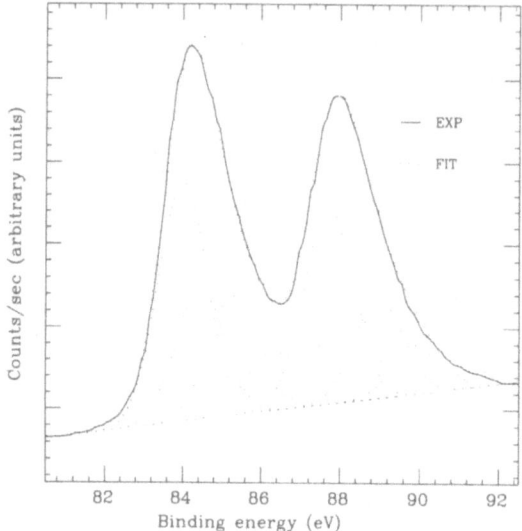

Fig. 9. The $Au4f_{7/2}$ and $Au4f_{5/2}$ peaks [48], with the "four-site" fit, as explained in the text. In increasing order of binding energy, the four spectral lines are from the bare surface sites, the core sites, the PPh_3-coordinated sites, and the Cl-coordinated sites

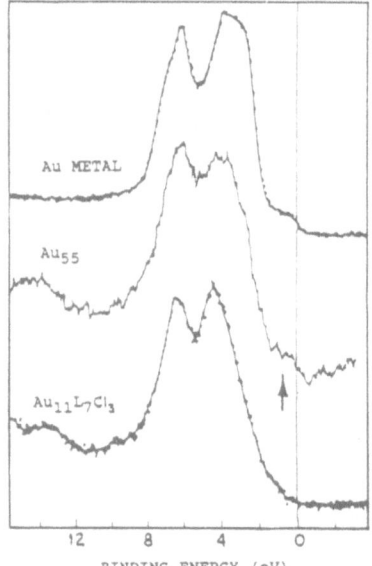

Fig. 10. A sample XPS valence-band spectrum of Au_{55} [48], sandwiched between a spectrum for bulk gold and for an $Au_{11}L_7X_3$ cluster. The latter two spectra were taken from Fig. 2 of Ref. [76]. The *arrow* denotes the position of the Fermi edge

Different series of spectra were obtained by single scans requiring about 5 to 10 minutes at a low beam power (120 W).

A second difference in the $Au5d$ peak between Quinten's [47] results and ours lies in the lack of any sign of the presence of ligands on the high binding energy side of the valence band spectra in Quinten's paper, while they are clearly visible on the left hand side of the two spectra shown in Fig. 10 for $Au_{11}L_7X_3$ and Au_{55}. This could be an indication that, in the unspecified experimental conditions used by them, their material has suffered ligand loss.

A third experimental difference is the overall shape of the Au4f core level peaks. The peaks in reference [47] were much more symmetric than ours.

5.2 Analysis of the Au4f Level

Au_{55} presents the highest Au4f trough depth (TD) value (defined as the ratio of the peak height at the valley between the two spin-orbit components to the peak maximum) [153] found throughout the series of gold molecular clusters which have been investigated under the same experimental conditions [48, 153]. A large TD value is a consequence of a composite Au4f peak. For comparison, a TD value for bulk gold of 0.20 was obtained on our experimental apparatus.

A chemically meaningful attempt at curve-fitting the Au4f peak implies several preliminary considerations. Firstly, the relatively low Au4$f_{7/2}$ b.e. value presented by Au_{55}, taken as the mid-point of the full width at half maximum (FWHM) of the peak, calls for a peak component very close to charge neutrality. This component cannot be assigned to P- or Cl-bonded Au atoms, since a comparison with literature values for Au4$f_{7/2}$ b.e.s clearly assigns these two components to oxidized metal atoms. Therefore, either the 13 core Au atoms and/or the 24 unligated surface Au atoms must account for this component.

Secondly, the unbonded surface Au atoms are much more numerous than the core atoms. Lack of experimental observation of a shoulder on the low b.e. side of the Au4$f_{7/2}$ peak therefore argues for an assignment of the first component in the Au4f peak to the surface Au atoms. Accordingly, as may be seen in Table 2, in the presently proposed fitting, the lowest b.e. component of the peak is assigned to the unbonded gold surface atoms.

Thirdly, there is no possible way to fit the peak component due to the 13 core atoms by locating this component on the high b.e. side of the Au4$f_{7/2}$ peak (See the procedure in ref. [76]), because of the shape of the peak. The highest b.e. component must be assigned to a component representing a relatively small number of Au atoms, i.e. those coordinated by chlorine.

Fourthly, as a consequence of the previous point, the Au_{55} cluster reveals a very different core than lower nuclearity gold molecular clusters such as $[Au_9L_8]^{3+}$ and $Au_{11}L_7X_3$, which all possess largely positively charged cores [76, 153]. In Au_{55} the core atoms have a relatively low b.e., showing a low oxidation state.

This last point suggests that the separation in Au4$f_{7/2}$ b.e. proposed earlier for P- and Cl-bonded gold atoms in $Au_{11}L_7X_3$, [76, 153] should be re-examined in the case of Au_{55}. The absence of a charged core in the metal frame should result in a larger separation of these two peak components, since in $Au_{11}L_7X_3$, the Cl-bonded Au atoms share their positive charge with the central gold atom [76].

The curve-fitting of the Au4f peaks in Fig. 9 has been obtained with the CERN library program MINUIT by making use of the above arguments. A common value of the FWHM of the peak height was imposed in the fit, and a common ratio of the 5/2 to 7/2 Au4f peak components, as deduced from the Au4f peak in bulk gold. The separation in energy of these two components was set at 3.7 eV, as in bulk Au in the same experimental conditions.

The applied curve-fitting resulted in four components for our Au$_{55}$ cluster, located at 84.0 eV, 84.4 eV, 85.3 eV, and 86.3 eV. These are assigned to the bare surface sites, the core sites, the PPh$_3$-coordinated sites, and the Cl-coordinated sites, respectively, with relative ratios reproducing the expected values from the Au$_{55}$ stoichiometry and structure (Table 2).

The proposed fit is obviously not the only possible way of curve-fitting the peak; it is presented as a part of an apparently consistent picture of the electronic structure of the very complicated species which the Au$_{55}$ cluster is. Statistically, however, the four-component fitting of our data is definitely superior to a two-component one, of the type reported by Quinten [47].

It is noteworthy that a similar small, albeit measurable difference in energy between surface and core sites is displayed by atoms near a flat gold surface [65]; this difference amounts to 0.4 eV, with the shallower component being assigned to surface atoms.

The b.e. value obtained for the core Au4f component in the present investigation being larger than in bulk gold is consistent with the MES I.S. of bulk gold and Au$_{55}$ and with the results from EXAFS experiments [39, 40, 41], which showed a smaller inter-gold distance than in the bulk. Under certain simplifying assumptions, i.e. ignoring the effects of possibly incomplete metallic screening, of 6s–5d hybridization, and of ionic charge, the b.e. shift can be related to the effective charge residing on the individual gold atoms. For scaling, Wertheim et al. [76] used the difference in binding energy between the average monomer value for PPh$_3$AuX (with X = Cl and I) and that for bulk gold, i.e. 1.4 eV, as being approximately equal to the loss of one 6s electron from the central gold atom in Au$_{11}$L$_7$X$_3$ compounds.

Because of the relative peak intensities and Au–P and AuCl shifts [76, 153], we know that the smallest peak must correspond to Cl-coordinated gold. The energy difference of 1.9 eV between our Au core site peak and the peak due to Cl-coordinated Au should then correspond to the loss of one 6s electron by the gold cluster to the chlorine atom; with nearly complete metallic screening, the negative charge on the chlorine will then be about one. It is this negative charge in the near vicinity of the ligated gold which leads to the observed core site b.e. shift. The b.e. shift between the two lowest kinetic energy peaks is 1.0 eV, which would then correspond to the loss of a bit more than one half 6s electron by the cluster to each phosphorus atom in the PPh$_3$-ligands. It is with this argument that we have estimated a reduction of the electron count of the cluster core by about 12. Should the actual charge on the chlorine atom be some fraction less than one, the charge on the phosphorus atom will differ accordingly.

It is also this proposed transfer of charge from the core of Au$_{55}$ to the ligands and the concomitant decrease in electronic repulsion which we feel is the reason for the difference in the interatomic distance of the core of Au$_{55}$ and that found for bare gold clusters of the same average diameter [83], which was mentioned in Sect. 3.5.

In Quinten's paper [47] the reported Au4f curve fitting into an overall ligated and unligated surface component and a core component is in strong contradiction with earlier XPS results for $Au_{11}L_7X_3$ [76] showing relevant chemical shifts from bare gold to Au–P and Au–Cl sites. The result of the Au4f curve fitting by Quinten gives a ratio of 1.70 between surface layer and core components. This in only half the expected ratio of $42/13 = 3.23$. In fact, their value is very close to the ratio of bare surface to core sites of $24/13 = 1.85$.

5.3 The Valence Band Results Compared to XPS on Other Systems

A standard technique for the comparison of XPS results on bare gold clusters supported on substrates [74, 75] is to relate surface coverage during deposition to average diameter of the particles formed by surface diffusion after deposition.

According to this, particles of 55 atoms correspond to a surface coverage just above 10^{15} atoms/cm^2, enabling us to place our XPS data from Au_{55} onto the plot from Ref. [74], since our Au_{55} clusters are covered by a jacket of non-conducting ligands and we know both the average number of gold nearest neighbors and the diameter of the metal core of Au_{55}. This is shown in Fig. 11.

There is quite satisfactory agreement between our data and that for the bare gold clusters supported on a non-conducting substrate, supporting the idea that

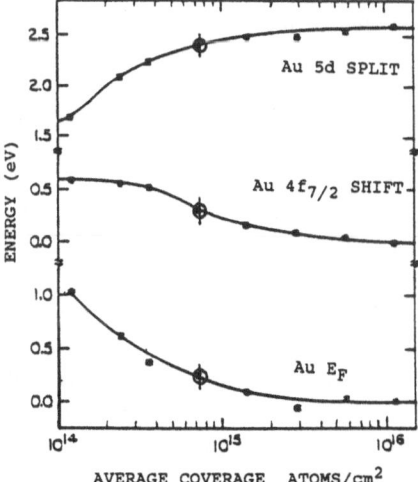

Fig. 11. A plot of the Au5d valence band splitting, of the Au4f$_{7/2}$ core level shift, and of the position of the Fermi level relative to the 5d band as a function of the coverage, taken from Fig. 2 of Ref. [74]. Our parameter values for Au_{55} have also been added to this plot as *crossed circles*

Au_{55} cluster surrounded by non-conducting ligands do resemble bare gold clusters of the same size.

The supported bare gold clusters [74] show a gradual and continuous transition from the $5d$ band shape exhibited by molecular gold clusters [76] to that of bulk gold. Even for very small clusters, there seems to be a clear development of a band structure. In the sense that the "band structure" of our cluster compound lies close to the bulk gold end of this range, we can say that the atoms in Au_{55} certainly appear to have metallic binding.

The experimentally observed separation between the $5d_{5/2}$ and $5d_{3/2}$ components of Au_{55} is 2.4 eV, closely approaching the bulk value, and is at variance with the smaller molecular gold clusters, where separations no higher than 1.9 eV between the same levels have been reported [76, 153]. A superposition of several slightly shifted doublets would inevitably broaden the two spin-orbit components and reduce the observed splitting. The high experimental value obtained for Au_{55} is in favour of a very intense contribution from a single component.

5.4   Comparison of XPS to MES and EXAFS Results

It is well established [154] that MES probes the initial electronic structure while photoemission samples the final state, after ejection of the electron. Both MES I.S. and XPS b.e. shifts are sensitive to the local chemical environment of the atom [155, 156]. However, the I.S. reflects the electronic charge density, of predominantly s- character, within the nuclear volume, while XPS b.e. shifts are a measure of the overall electronic charge density from all the shells.

One can think of I.S. changes in terms of s-electron density as a first approximation. The amount of charge needed to give the decrease of nearly 0.9 mm/s in the I.S. of the 13 core sites relative to bulk gold (with the 10% volume contracted situation taken as a reference value), using the value of the volume coefficient of the I.S. $\lambda_v = -6.8$ mm/s for gold [126], should lead to an average increase of about 0.3 mm/s for the 42 surface sites. On the other hand, in our application of the model of Citrin and Wertheim [65], use has been made of the idea that the surface sites, due to their lower coordination, have more $5d$-like character than the bulk (or core) sites. This would mean that the charge which has left the core sites of the cluster in order to explain their I.S. value is also partially converted into $5d$-like electrons on the surface. This would also mean a small decrease in the surface I.S. values, rather than an increase, again in agreement with observation.

Our XPS results on Au_{55} can also be examined to decide whether metallic shielding is present. The presence of a finite density of states at the Fermi level in Au_{55} was clearly detected in our XPS valence band spectrum, as indicated by the arrow in Fig. 10. This presence can be considered as an indication of metallic character in a cluster, even though this view has been questioned [74, 152, 157]. In addition, the near full bulk value of the valence band splitting of Au_{55} is also a positive indication. Furthermore, the Au4f bare surface and core components have been experimentally resolved by the applied curve-fitting procedure, and as expected from the lineshape observed, the bare surface contribution lies at the lowest b.e., with the core component following closely in energy.

These results, taken together, give a strong indication that all the atoms of the gold core of the Au_{55} cluster compound are already well into the state of metallic binding, i.e. that the valence electrons have become delocalized, and that metallic screening in this system can be considered to be nearly complete.

Looking at the structure of Au_{55} in comparison to lower nuclearity clusters such as $Au_9L_8^{3+}$ and $Au_{11}L_7X_3$ one might wonder whether the transition to metallic behavior is due to the nuclearity of the cluster or to the fact that the Au_{55} cluster has assumed a cubic close-packed structure. The EXAFS results of Balerna et al. [83, 84] mentioned above, which indicate fcc structure even for an average particle diameter of only 1.1 nm, may be an indication that it is indeed the ligand surroundings for the smaller clusters, together with electron counting procedures [11, 12, 13], which determine their icosahedron-based cluster structure. In higher-nuclearity clusters, as a metallic band structure begins to develop, the close link between cluster geometry and electron count is lost [12].

The low nuclearity gold clusters show decided directionality in their bonding, which is dominated by radial covalent bonding to the central atom, with weaker tangential interaction on the cluster surface. This also holds for clusters with open crown structures such as $[Au_9\{P(p\text{-}C_6H_4OMe)_3\}_8]^{3+}$ [14], which have longer tangential than radial Au–Au distances although they are not forced to do so by icosahedral geometry. In Au_{55} on the other hand, the bonding in the cluster is much more non-directional. The cluster $[Au_{39}(PPh_3)_{14}Cl_6]^{2+}$ [38] may represent an intermediate stage in the transition between the two structural and bonding types.

A number of years ago, Lee et al. [117] reported XPS measurements on supported gold clusters, using variable photon energies. From the development of the shape of the valence band as a function of cluster size and measured as a function of energy, they concluded that the average number of gold atoms needed to approach metallic behavior must be about 200. This is not in conflict with our conclusion of metallic binding, since their work is concerned more with the development of the crystalline momentum k as a valid quantum number than with the delocalization of the valence electrons.

This again illustrates the need for care in defining the criteria one is using for "metallic behavior".

6 Conclusions

In conclusion, we can with confidence make a number of statements about the cluster compound Au_{55}:

Almost all of the experimental evidence is that the gold core of this material is indeed based on a cubic close-packed, cuboctahedral stacking of gold atoms [16, 26]. The single inter-gold distance derived from the EXAFS measurements, and the number of Mössbauer sub-spectra observed, rule out an icosahedral structure. The excellent agreement between the relative f-factors found in the Mössbauer measurements as compared to those obtained from the model calculations, require a three layer magic number structure, with a total gold core mass of the order of 55 gold atomic masses.

Measurement of the decomposition enthalpy demonstrates the stronger binding in the cluster compound relative to the bulk, in agreement with the shorter gold-gold distance observed by EXAFS.

From the Mössbauer results on Au_{55}, the thermodynamic behavior of Au_{55}, as well as $Au_9L_8^{3+}$ and $Au_{11}L_7(SCN)_3$, at temperatures between 2 K and 30 K can be understood on the basis of the center-of-mass motion of the whole gold core of the material, within the matrix formed by the surrounding ligands. Furthermore, none of these clusters are in the liquid (fluxional) state on the MES time scale, up to at least 30 K.

The coordination numbers based on this structure work extremely well for describing the microscopic physical properties of this material, including the Mössbauer I.S.s of the surface sites and of the specific heat of the clusters below about 65 K. No linear electronic term in the specific heat is seen down to 60 mK, due to the still significant T^3 contribution from the center-of-mass motion still present at this temperature. The Schottky tail which develops below 300 mK in magnetic fields above 0.4 T has been quantitatively explained by nuclear quadrupole contributions.

The AC and DC electrical conductivity results could be well described by standard models for variable range hopping between nearest neighbors in an array of identical conducting spheres isolated in a non-conductive medium at temperatures above 77 K. This is also in good agreement with the results obtained above, but requires consideration of the intercluster interactions as well. It also gives an indication that below 77 K correlated electronic motion is strongly hindered.

Although the NMR results do not appear to be at variance with the other microscopic measurement techniques, they do not seem to provide any real indication as to whether or not the Au_{55} cluster material is metallic.

The behavior of the magnetic susceptibility as a function of temperature is in accord with what one should expect from an array of small isolated metal clusters, with splitting between the filled and the empty intra-cluster electronic energy bands of the order of 30 K. The magnetic moment per cluster of less than one electron spin still requires clarification.

ESR results on both Au_{55} and on gold colloids are ambiguous, and it is clear that additional experimental efforts on both systems are necessary.

The position and shape of the X-ray absorption edge of Au_{55} indicates collective metallic behavior at room temperature.

The lack of a clearly developed peak in the UV-visible absorption spectrum due to plasma resonance places a limit on the collective metallic behavior of the electrons in the cluster. On the other hand, the $5d \rightarrow 6s,6p$ interband absorption is well developed toward that of large colloidal gold particles.

The MES I.S. values can be well understood on the basis of metallic binding by the electrons in the cluster core.

The XPS $Au4f$ core level b.e.s of Au_{55} can be semi-quantitatively understood. The $Au4f$ core level spectrum can be well fitted with a four-site model with site populations corresponding to those expected for a cuboctahedron, in good agreement with the Mössbauer I.S. results. MES and XPS measurements consistently show that the core metal atoms in $Au_{11}L_7X_3$ clusters bear a significant positive charge, but that this charge is absent in Au_{55}.

The shape and splitting of the XPS $Au5d$ band in the 55 atom cluster material, with its insulating jacket of ligands, reproduces nicely the basic shape of the $5d$ band of bulk gold, including a clearly visible density of states at the Fermi level.

To sum up all the experimental evidence: whereas $Au_{11}L_7X_3$ and smaller molecular cluster compounds are covalently bonded and show no tendency toward metallic binding, the bonding in $Au_{55}(PPh_3)_{12}Cl_6$ is delocalized, nondirectional, and substantially metallic in character.

Acknowledgements. We are deeply indebted to G. Schmid whose generosity in providing samples for most of the measurements discussed here made this work possible. A number of colleagues have also contributed in large measure to this manuscript, both through lively discussions, as well as through their constructive comments. We would especially like to thank H. B. Brom, J. van Ruitenbeek, L. J. de Jongh, S. Mukhin, F. Mulder, R. J. Newport, J. A. Creighton, Z. Stadnik, and D. Rancourt.

This work has been supported by several different organizations: the Foundation for the Fundamental Research on Matter (FOM), made possible by financial support from the Netherlands Organization for the Advancement of Pure Research (ZWO); the Royal Society; the United Kingdom Science & Engineering Research Council (SERC); and the Italian National Research Council (CNR).

The international cooperation for this research was made possible by the EG project ST2J 0084.

References

1. Vollenbroek FA (1979) PhD. Thesis Univ. of Nijmegen, The Netherlands
2. Vollenbroek FA, van der Velden JWA, Bour JJ, Trooster JM (1981) Rec J Royal Neth Chem Soc 100: 375

3. Vollenbroek FA, Bouten PCP, Trooster JM, van den Berg JP & Bour JJ (1978) Inorg Chem 17: 1345
4. Vollenbroek FA, Bour JJ, Trooster JM, van der Velden JWA (1978) J Chem S Ch 907
5. Bellon PL, Manassero M & Sansoni M (1972) J Chem S Da 1481
6. Albano VG, Bellon PL, Manassero M & Sansoni M (1970) J Chem S Ch 1210
7. van der Velden JWA (1983) PhD. Thesis Univ. of Nijmegen, The Netherlands
8. v.d.Velden JWA, Vollenbroek FA, Bour JJ, Beurskens PT, Smits JMM & Bosman WP (1981) J Roy Neth Chem Soc 100: 148
9. Bos W, Kanters RPF, van Halen CJ, Bosman WP, Behm H, Smits JMM, Beurskens PT, Bour JJ & Pignolet LH (1986) J Orgmet Ch 307: 385
10. Longoni G (1990) Pure & Applied Chemistry 62: 1183
11. Mingos DMP (1976) J Chem S Da 1163
12. Mingos DMP (1985) J Chem S Ch 1352
13. Mingos DMP, Slee T & Zhenyang L (1990) Chem Rev 90: 383
14. Hall KP, Theobald BRC, Gilmour DI, Welch AJ & Mingos DMP (1982) J Chem S Ch 528
15. Hall KP & Mingos DMP (1984) Prog Inorg 32: 237
16. Schmid G (1985) Structure and Bonding 62: 51
17. Benfield RE (1989) J Orgmet Ch 372: 163
18. Benfield RE (1989) Z Phys D12: 453
19. Pronk BJ, Brom HB, de Jongh LJ, Longoni G & Ceriotti A (1986) Sol St Comm 59: 349
20. Pronk BJ, Brom HB, Ceriotti A & Longoni G (1987) Sol St Comm 64: 7
21. de Jongh LJ, Brom HB, Longoni G, Nugteren PR, Pronk BJ, Schmid G, Smit HHA, van Staveren MPJ, Thiel RC (1987) In: Jena P, Rao BK, Khanna SN (eds) *Physics and Chemistry of Small Clusters*, Plenum, London, p 807
22. de Jongh LJ, Albino O de Aguiar J, Brom HB, Longoni G, van Ruitenbeek JM, Schmid G, Smit HHA, van Staveren MPJ & Thiel RC (1989) Z Phys D12: 445
23. de Jongh LJ, Baak J, Brom HB, van der Putten D, van Ruitenbeek JM & Thiel RC (1991) Proc Intl Symposium on the Physics and Chemistry of Finite Systems: from Clusters to Crystals; Richmond, VA
24. Smit HHA (1998) PhD. Thesis Univ. of Leiden, The Netherlands
25. Smit HHA, Nugteren PR, Thiel RC & de Jongh LJ (1988) Physica B153: 33
26. Schmid G, Pfeil R, Boese R, Bandermann F, Meyer S, Callis GHM & vd Velden JWA (1981) Chem Ber 114: 3634
27. Schmid G, Klein N, Korste L, Kreibig U & Schönaure D (1988) Polyhedron 7: 605
28. The methylated version of Au_{55} has been prepared by G. Schmid; ^{31}P NMR [29] and electrical conductivity [29, 30] results have been reported
29. van Staveren MPJ (1989) PhD. Thesis Univ. of Leiden, The Netherlands
30. van Staveren MPJ, Brom HB & de Jongh LJ (1991) Physics Reports 208: 1–96.
31. Schmid G, Giebel U, Huster W & Schwenk A (1984) Inorg Chim 85: 97
32. Schmid G (1988) Polyhedron 7: 2321
33. Fackler JP Jr, McNeal CJ, Winpenny REP & Pignolet LH (1989) J Am Chem S 111: 6434
34. Teo BK, Hong MC, Zhang H & Huang DB (1987) Angew Chem Intl Edn 26: 897
35. Feld H, Leute A, Rading D, Benninghoven A & Schmid G (1990) Z Phys D17: 73
36. Feld H, Leute A, Rading D, Benninghoven A & Schmid G (1990) J Am Chem S 112: 8166
37. Teo BK & Zhang H (1990) J Cluster Science 1: 155
38. Teo BK, Shi X & Zhang H (1992) J Am Chem S 114: 2743
39. Fairbanks MC, Benfield RE, Newport RJ, & Schmid G (1990) Sol St Comm 74: 431
40. Marcus MA, Andrews MP, Zegenhagen J, Bommannavar AS & Montano P (1990) Phys Rev B42: 3312
41. Cluskey PD, Newport RJ, Benfield RE, Gurman SJ & Schmid G Materials Research Society Spring 1992 Meeting Proceedings (submitted for publication)
42. Smit HHA, Thiel RC, de Jongh LJ, Schmid G & Klein N (1988) Sol St Comm 65: 915
43. Kramer P, Dirken MW & Thiel RC (unpublished calculations)
44. Thiel RC, Dirken MW & Zanoni R (1990) Proceedings ICAME Budapest '89: Applications of the Mössbauer Effect, J.C. Baltzer, Bazel 1729
45. Mulder FJ, Thiel RC & van der Zeeuw EA (unpublished MES results on an absorber produced by dissolving Au_{55} in CH_2Cl_2 and evaporating solution)
46. Mulder FJ, van der Zeeuw EA & Thiel RC (1992) Sol St Comm (submitted for publication)
47. Quinten M, Sander I, Steiner P, Kreibig U, Fauth K & Schmid G (1991) Z Phys D20: 377
48. Zanoni R (to be published)

49. de Jongh LJ, Brom HB, van Ruitenbeek JM, Thiel RC, Schmid G, Longoni G, Ceriotti A, Benfield RE, Zanoni R (in press) In: Pacchioni G, Bagus PS (eds) *Cluster Models for Surface and Bulk Phenomena* (NATO ASI Series B)
50. van Staveren MPJ, Brom HB, de Jongh LJ & Schmid G (1991) Z Phys D20: 333
51. van Staveren MPJ, Brom HB, de Jongh LJ & Schmid G (1989) Z Phys D12: 451
52. Benfield RE (unpublished work)
53. van Ruitenbeek JM (unpublished work)
54. Goll G, Löhneysen Hv, Kreibig V & Schmid G (1991) Z Phys D20: 329
55. Gmelin E (private communication)
56. Baak J (private communication)
57. Benfield RE, Creighton JA, Eadon DG & Schmid G (1989) Z Phys D12: 533
58. Benfield RE & O'Brien P (unpublished work)
59. Fauth K, Kreibig U & Schmid G (1989) Z Phys D12: 515
60. Fauth K, Kreibig U & Schmid G (1991) Z Phys D20: 297
61. van Staveren MPJ, Brom HB, de Jongh LJ & Schmid G (1986) Sol St Comm 60: 319
62. van Staveren MPJ, Moonen JT, Brom HB, de Jongh LJ & Schmid G (1989) Z Phys D12: 461
63. Brom HB, van Staveren MPJ & de Jongh LJ (1991) Z Phys D20: 281
64. Mees A (1987) Masters Thesis, U. of Leiden, The Netherlands (unpublished)
65. Citrin PH & Wertheim GK (1983) Phys Rev B27: 3176
66. Miedema AR & Woude Fvd (1980) Physica 100B: 145
67. Thiel RC & Dirken MW: unpublished calculations of the MES I.S.s of the surface sites of Au_{55}.
68. Kreibig U, Fauth K, Granqvist C.-G & Schmid G (1990) Z Phys Chemie Neue Folge 169: 11
69. Kreibig U (1974) J Phys Met Phys 4: 999
70. Kreibig U & Genzel L (1985) Surf Sci 156: 678
71. Kreibig U (1986) In: Davenas D, Rabette PM (eds) *Contribution of Cluster Physics to Materials Science and Technology*, Nijhoff, Den Haag, 373
72. Creighton JA & Eadon DG (1991) J Chem S F 87: 3881
73. Kreibig U, Althoff A & Pressmann H (1981) Surf Sci 106: 308
74. Wertheim GK, Di Cenzo SB & Youngquist SE (1983) Phys Rev Lett 51: 2310
75. Mason MG (1983) Phys Rev B27: 748
76. Wertheim GK, Kwo J, Boon K Teo & Keating KA (1985) Sol St Comm 55: 357
77. Wallenberg LR, Bovin JO & Schmid G (1985) Surf Sci 156: 256
78. Wierenga HA, Soethout L, Gerritsen JW, van der Leemput BEC, van Kempen H & Schmid G (1990) Advanced Materials 2: 482
79. van der Leemput LEC, Gerritsen JW, Rongen PHH, Smokers RTM, Wierenga HA, van Kempen H & Schmid G (1991) J Vac Sci Technology B9: 814
80. Becker C, Fries Th, Wandelt K, Kreibig U & Schmid G (1990) J Vac Sci Technology B9: 810
81. Mackay AL (1962) Act Cryst 15: 916
82. Briant CE, Theobald BRC, White JW, Bell LK, Mingos DMP & Welch AJ (1981) J Chem S Ch 201
83. Balerna A, Bernieri E, Picozzi P, Reale A, Santucci S, Burattini E & Mobilio S (1985) Phys Rev B31: 5058
84. Balerna A, Bernieri E, Picozzi P, Reale A, Santucci S, Burattini E & Mobilio S (1985) Surf Sci 156: 206
85. Cariati F & Naldini L (1971) Inorg Chim Acta 5: 172
86. Cariati F & Naldini L (1972) J Chem S D 2286
87. Housecroft CE, Wade K & Smith BC (1978) J Chem S Ch 765
88. Ames LL & Barrow RF (1967) Trans Faraday Soc 63: 39
89. Gingerich KA (1980) Faraday Symp Chem Soc 14: 109
90. Schmidbaur H, Dwizok K, Grohmann A & Müller G (1989) Chem Ber 122: 893
91. Parish RV, Moore LS, Dens AJJ, Mingos DMP & Sherman DJ (1989) J Chem S 539
92. Thiel RC (1967) Z Phys 200: 227
93. Marshall SW & Wilenzick RM (1966) Phys Rev Letters 16: 219
94. Balerna A & Mobilio S (1986) Phys Rev B34: 2293
95. van Wieringen JS (1968) Phys Lett A26: 370
96. Viegers MPA (1976) PhD. Thesis Univ. of Nijmegen, The Netherlands
97. Viegers MPA & Trooster JM (1977) Phys Rev B15: 72
98. van Leeuwen DA (unpublished results)
99. Baak J & Brom HB (1991) Proc. Intl. Symposium on the Physics and Chemistry of Finite Systems: from Clusters to Crystals; Richmond, VA

100. Sette F, Chen CT, Rowe JE & Citrin PH (1987) Phys Rev Lett A26: 311
101. Johnson BFG & Benfield RE (1978) J Chem S D 1554
102. Sawada S & Sugano S (1991) Z Phys D20: 259
103. Kostelitz M & Domange JL (1973) Sol St Comm 13: 241
104. Couchmann PR & Ryan CL (1978) Phil Mag 37: 369
105. Solliard C & Flueli M (1985) Surf Sci 156: 487
106. Solliard C (1984) Sol St Comm 51: 947
107. Buffat Ph & Borel J-P (1976) Phys Rev A13: 2287
108. Couchmann PR (1979) Phil Mag A40: 637
109. Labastie P & Whetten RL (1990) Phys Rev Lett 65: 1567
110. Iijima S & Ichihashi T (1986) Phys Rev Lett 56: 616
111. Garzon IL & Jellinek J (1991) Z Phys D20: 235
112. Benfield RE, Braga D & Johnson BFG (1988) Polyhedron 7: 2549
113. Blume M & Tjon JA (1968) Phys Rev 165: 446
114. Smit HHA & Thiel RC (unpublished MES measurements on gold cluster-metal [32])
115. Schmid G, Klein N (1987) In: Jena P, Rao BK, Khanna SN (eds) *Physics and Chemistry of Small Clusters*, Plenum, London, 329
116. Kittel C (1976) *Introduction to Solid State Physics*, 5th edn, J Wiley, New York, p 154
117. Lee ST, Apai G, Mason MG, Benbow R & Hurych Z (1981) Phys Rev B23: 505
118. Meyer JM & Allred AL (1968) J Inorg Nucl Chem 30: 1328
119. Strohmeier W & Müller FJ (1967) Chem Ber 100: 2812
120. Rademann K, Kaiser B, Even U & Hensel F (1987) Phys Rev Lett 59: 2319
121. Rademann K (1989) Ber Bunsenges Phys Chem 93: 653
122. Kaiser B & Rademann K (1991) Z Phys D19: 227
123. Haberland H, Kornmeier H, Langosch H, Oschwald M & Tanner G (1990) J Chem S F 86: 2473
124. Bennemann KH (1990) Z Phys Chem 169: 29
125. Shenoy GK, Wagner FE (eds) (1978) Mössbauer isomer shifts, North-Holland, Amsterdam
126. Williamson DL, In: Shenoy GK, Wagner FE (eds) Mössbauer Isomer Shifts, North-Holland, Amsterdam, chap 6b
127. Citrin PH, Wertheim GK & Baer Y (1983) Phys Rev B27: 3160
128. van der Woude F & Miedema AR (1981) Sol St Comm 39: 1097
129. Miedema AR, de Boer FR & de Chatel PF (1973) J Phys F3: 1558
130. Pettifor DG (1979) Phys Rev Lett 42: 846
131. Varma CM (1979) Sol St Comm 31: 295
132. Williams AR, Gallet CD & Moruzzi VL (1980) Phys Rev Lett 44: 429
133. de Graaf H (1982) PhD Thesis, Univ. of Leiden, The Netherlands
134. de Vries JWC (1984) Ph.D. Thesis, Univ. of Leiden, The Netherlands
135. Summerfield S & Butcher PN (1982) J Phys C15: 7003
136. Summerfield S & Butcher PN (1983) J Phys C16: 295
137. Miller A & Abrahams E (1960) Phys Rev 120: 745
138. Summerfield S (1985) Philos Mag 52: 9
139. van Ruitenbeek JM (1991) Z Phys D 19: 247
140. van Ruitenbeek JM & van Leeuwen DA (1991) Phys Rev Lett 67: 640
141. van Ruitenbeek JM: private communication.
142. Kittel C (1976) *Introduction to Solid State Physics*, 5th edn, J Wiley, New York, p 167
143. Kittel C (1976) *Introduction to Solid State Physics*, 5th edition, J Wiley & Sons, New York p 453
144. Brom HB, Baak J, de Jongh LJ, Mulder FM, Thiel RC & Schmid G (1992) Z Phys D (accepted for publication) & Baak J, Brom HB, de Jongh LJ & Schmid G (1992) Z Phys D (accepted for publication)
145. Schober C, Kurz G, Wonn H, Nemoshkalenko VV & Antonov VN (1986) Physica Status Solidi B136: 233
146. Schober C & Antonov VN (1987) Physica Status Solidi B143: K31
147. Dupree R, Forwood CT & Smith MJA (1967) Physica Status Solidi 24: 525
148. Monot R, Chatelain A & Borel JP (1971) Phys Lett A34: 57
149. Monod P & Janossy A (1977) Low-temperature Physics 26: 311
150. Elder RC & Eidsness MK (1987) Chem Rev 87: 1027
151. Kerker M (1969) The scattering of light and other electromagnetic radiation, Academic, New York, chap 3

152. Mason MG (1984) in *Cluster Models for Surface and Bulk Phenomena* (NATO ASI Series B, ed. Pacchioni G. & Bagus P. S.) (in press)
153. Battistoni C, Mattogno G, Zanoni R & Naldini L (1982) J Electr Spectr 28: 32
154. Gelius U (1974) Phys Scripta 9: 133
155. Nagy DL, Lázár K & Kajcsos Zs (1989) Applications of the Mössbauer Effect, Proc Intern Conf Appl Möss Spectr JC Baltzer, Bazel
156. Thosar BV, Iyengar PK, Bhargava SC & Srivastava JK (1983) *Advances in Mössbauer Spectroscopy*, Studies in Physical and Theoretical Chemistry Elsevier, Amsterdam
157. Di Cenzo SB & Wertheim GK (1985) Comments Solid State Phys 11: 203

Recent Aspects of the Structure and Reactivity of Cyclophosphazenes

V. Chandrasekhar and K. R. Justin Thomas

Department of Chemistry, Indian Institute of Technology, Kanpur-208016, India

This review describes the reactivity and structural aspects of cyclophosphazenes. The reactions of halogenocyclophosphazenes with a variety of monofunctional and difunctional reagents are described. The mechanisms of the reactions involved are discussed in light of kinetic evidence and product analysis. Reactions involving main-group and transition-metal reagents and halogenocyclophosphazenes are also described. The chemistry of hydridophosphazenes containing a P–H bond is described in more detail. Some of the rearrangement reactions including *cis-trans* isomerism and tautomerism which are common to some cyclophosphazene derivatives have been discussed. The coordination chemistry involving cyclophosphazenes as ligands via endocyclic and/or exocyclic coordination sites resulting in several types of derivatives including metallocenyl cyclophosphazenes, metal clusters and carboranyl cyclophosphazenes has been critically evaluated. The structural aspects of cyclophosphazenes have been reviewed with particular emphasis on multinuclear NMR and single crystal X-ray methods. Bond angle-bond length variations in X-ray structures of cyclophosphazenes have been explained based on the theories of bonding prevalent in these molecules.

1 Introduction 43
2 Preparative Methods 45
3 Reactions with Amines 48
  3.1 Reactions of $N_3P_3Cl_6$ with Amines 48
  3.2 Reactions of $N_4P_4Cl_8$ with Amines and the Formation of Bicyclic Phosphazenes . 51
4 Reactions with Hydroxy Compounds and Mercaptans 56
5 Reactions with Difunctional Reagents 56
6 Isomerisation, Tautomerism and Thermal Rearrangements 60
  6.1 *cis* ↔ *trans* Isomerisation 60
  6.2 Tautomerisations (O ↔ N Hydrogen Migrations) 61
  6.3 Rearrangements of Alkoxy Cyclophosphazenes (O ↔ N Alkyl Migrations) 64
  6.4 Other Miscellaneous Rearrangements 66
7 Reactions of Main Group Organometallic Reagents with Cyclophosphazenes 67
  7.1 Reactions with Organolithium Reagents 68
  7.2 Reactions with Grignard Reagents 71
  7.3 Reactions of Organophosphazenes with Alkyl Lithium Reagents 73
  7.4 Friedel-Crafts Reactions 74
8 Hydridophosphazenes: Syntheses and Reactivity 76
9 Interaction with Transition Metal Ions 81
  9.1 Ring Nitrogen Coordination 81
  9.2 Exocyclic Group Coordination and Direct Ring Phosphorus Interaction with Transition Metals 82

10 Physical Methods   86
   10.1 Proton NMR   86
   10.2. ^{31}P NMR Spectroscopy   88
   10.3 Other Spectroscopic Techniques   93
   10.4 X-ray Diffraction Studies   94
11 Summary and Conclusions   104
12 References   106

1 Introduction

Cyclophosphazenes occupy a prominent place among the inorganic heterocyclic ring systems [1]. These ring compounds contain a $\left[-N=P-\right]_n$ repeating unit where nitrogen is trivalent and dicoordinate while phosphorous is pentavalent and tetracoordinate. Two exocyclic substituents are present on phosphorus while none are present on nitrogen. Although the ring size can vary from four (n = 2) to a maximum isolated ring size of thirty four (n = 17) the most widely studied examples are the six membered (n = 3) and eight memberd rings (n = 4) [1].

Research interest in this class of heterocyclic compounds arises in part due to the wide range of chemical reaction pathways observed in the substitution reactions involving the P–X bond in $N_3P_3X_6$ and $N_4P_4X_8$ (X = Cl, Br and F) by various types of nucleophiles including amines (aliphatic and aromatic), alcohols, phenols, thiols, main group and transition metal organometallic reagents etc. [2]. Several positional and geometric isomers are possible beyond the mono stage of substitution [3] (Fig. 1). Interest in unravelling the factors involved in the regio and stereochemical pathways observed with different nucleophiles has formed a major impetus for research on these group of compounds.

A second reason of interest in cyclophosphazenes has stemmed from the facile ring opening polymerization of the hexachlorocyclotriphosphazene, $N_3P_3Cl_6$, to the linear macromolecule $[NPCl_2]_x$ [4, 5] (Eq. 1).

$$N_3P_3Cl_6 \xrightarrow{\Delta} \left[-N=P-\right]_x \longrightarrow \left[-N=P-\right]_x \tag{1}$$
$$\text{x = 15000} \qquad \text{R = amino, alkoxy}$$
$$\text{(aryloxy) etc.}$$

The parent polymer by itself is not a useful material owing to the extreme hydrolytic sensitivity of the P–Cl bond. However, this feature has been turned around and used as an advantage. Nucleophilic substitution of the chlorines in the polymer results in substituted polyphosphazenes which are hydrolytically stable. Also, using this method the polymer architecture and properties are readily fine-tuned by a subtle variation of the substituent. Over three hundred types of polyphosphazenes have been synthesised by this method. Assembly of organic polymers containing cyclo-phosphazenes as pendant groups is another approach that is gaining importance [6].

A third feature which stimulates interest in cyclophosphazenes is their behaviour as ligands towards transition metals [7, 8]. Several types of coordination pattern involving either the skeletal ring nitrogen atoms and/or suitable exocyclic substituent donor atoms have been documented (Fig. 2). This area of research has been receiving widespread attention in recent years [7].

The last decade has seen the application of several state of the art instrumental methods for the structural elucidation of cyclophosphazenes in solution as

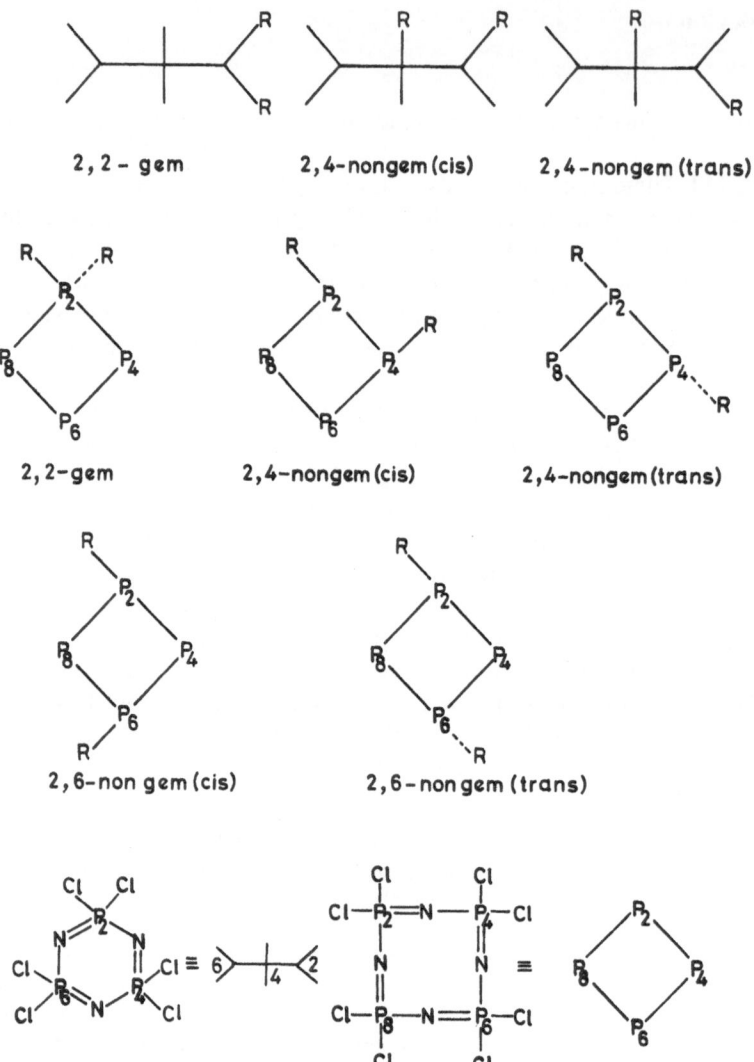

Fig. 1. Some of the positional and stereo isomers possible in the substitution reactions of $N_3P_3Cl_6$ and $N_4P_4Cl_8$. Numbering starts at nitrogen

well as in the solid state. There have been attempts at correlating the data obtained from different methods with the nature of bonding in these ring systems [9, 10]. Although several simple bonding models which allow an explanation of the observed structural parameters as well as the reaction pathways are now available [9], the exact nature of bonding and electronic structure in these systems has not yet been delineated and is still being investigated [11].

This article will review the current status of the structure and reactivity aspects of cyclophosphazenes. It is not an attempt to present a comprehensive

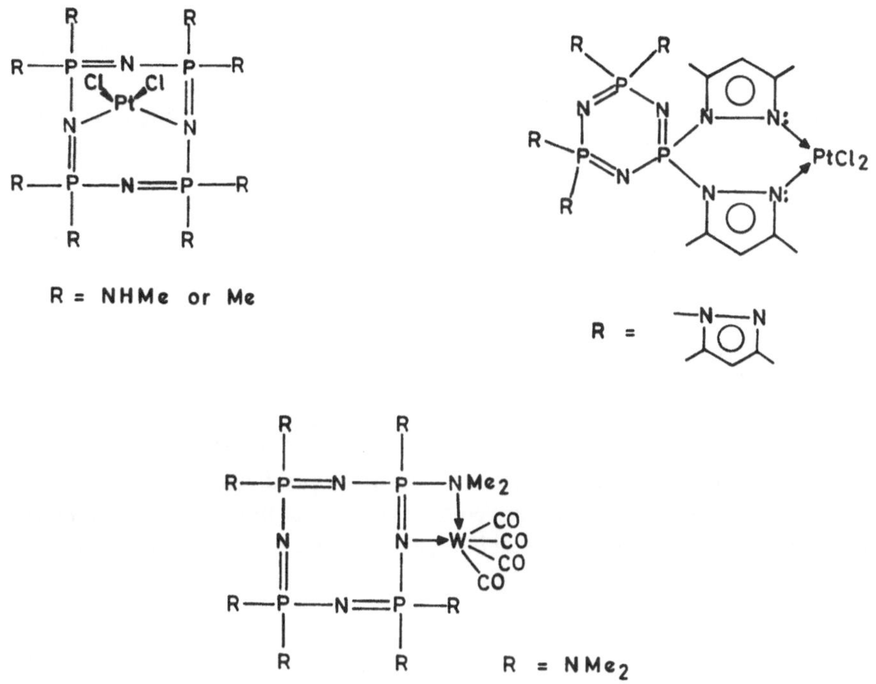

Fig. 2. Examples of cyclophosphazene coordination behaviour

survey. On the other hand some representative and well-studied systems will be reviewed. The nomenclature followed in this chapter is according to Shaw and coworkers [12]. Accordingly nitrogen is numbered first followed by phosphorus (Fig. 1) Trivial names such as trimeric chloride and tetrameric chloride are used for $N_3P_3Cl_6$ and $N_4P_4Cl_8$ respectively. Cyclophosphazenes have been reviewed before in the form of monographs, and summaries [1–3, 13]. Also the current literature is reviewed annually [14].

2 Preparative Methods

Chloro and bromo cyclophosphazenes are generally prepared by the reaction of chloro or bromophosphoranes with ammonium halides [15] (Eq. 2 and Scheme 1). Special methods are also available for the synthesis of chlorocyclophosphazenes (Scheme 1). In these reactions cyclic oligomers and linear products are formed, and the former compounds are easily separated from the mixture because of their greater solubility in petroleum solvents. Maximization of cyclic

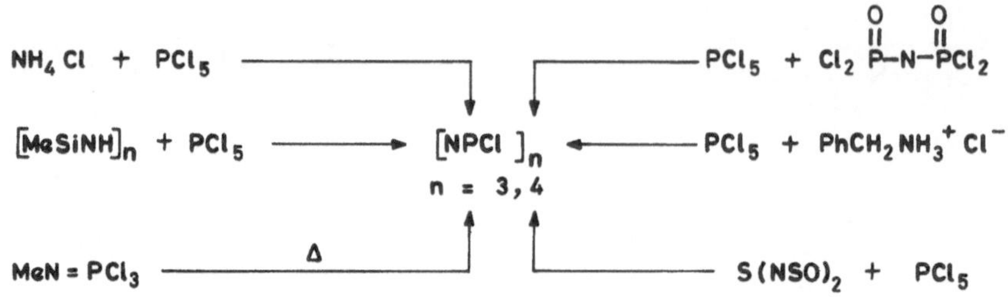

$$NH_4Cl + PCl_5 \xrightarrow{\qquad}$$

$$[MeSiNH]_n + PCl_5 \longrightarrow [NPCl]_n \longleftarrow$$
$$n = 3,4$$

$$PCl_5 + Cl_2\ \overset{O}{\underset{}{P}}\!\!-\!\!N\!\!-\!\!\overset{O}{\underset{}{P}}Cl_2$$

$$PCl_5 + PhCH_2NH_3^+\ Cl^-$$

$$MeN = PCl_3 \xrightarrow{\quad \Delta \quad}$$

$$S(NSO)_2 + PCl_5$$

Scheme 1

product formation was the subject of research for some time. Several catalysts have been developed [16].

$$NH_4Br \text{ or } NH_3 + PBr_3 + Br_2 \longrightarrow [NPBr_2]_{3,4} \tag{2}$$

Fluorocyclophosphazenes are isolated as main products in many reactions [17–19] (Eqs. 3–5). However, convenient synthesis of fluorocyclophosphazenes is usually achieved by metathetical halogen exchange reactions starting from the chloro analogues (Eq. 6) [20–22].

$$NF_3 + P_4S_3 + P_4S_{10} \xrightarrow[180-215°C]{\Delta} [NPF_2]_{3-9} \tag{3}$$

$$NF_3 + P_4S_3 \xrightarrow{700°C} [NPF_2]_{3,4} \tag{4}$$

$$NF_3 + \text{red } P \xrightarrow{\Delta} [NPF_2]_3 + \text{other products} \tag{5}$$

$$[NPCl_2]_n \xrightarrow[\substack{\text{or KSO}_2\text{F} \\ \text{or MF/18-Crown-6} \\ M = \text{alkali metal}}]{MF \text{ or } SbF_3} [NPF_2]_n \quad n = 3 \text{ or } 4 \tag{6}$$

Interesting methods have been developed for the synthesis of organophosphazenes [23]. These are summarized in Scheme 2. For example, the N–silyl–P–halo or alkoxyphosphoranimines upon decomposition afford organocyclophosphazenes [24]. Other miscellaneous reactions employed for the synthesis of organocyclophosphazenes are enumerated in Eqs. (7–11).

$$[RR'P(NH_2)_2]^+Cl^- + (PPh_2)_2NH \xrightarrow[Et_3N]{CCl_4} gem\text{-}N_3P_3Ph_4RR' \tag{7}$$

$$R = R' = Me \text{ or } C_6H_{11}; \ R = Me, R' = Et \text{ or } Ph$$

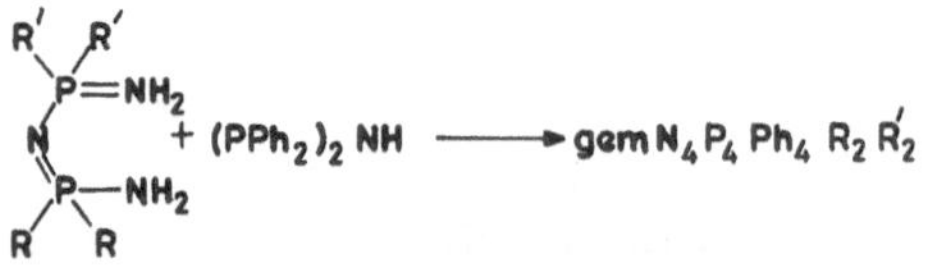

$$\tag{8}$$

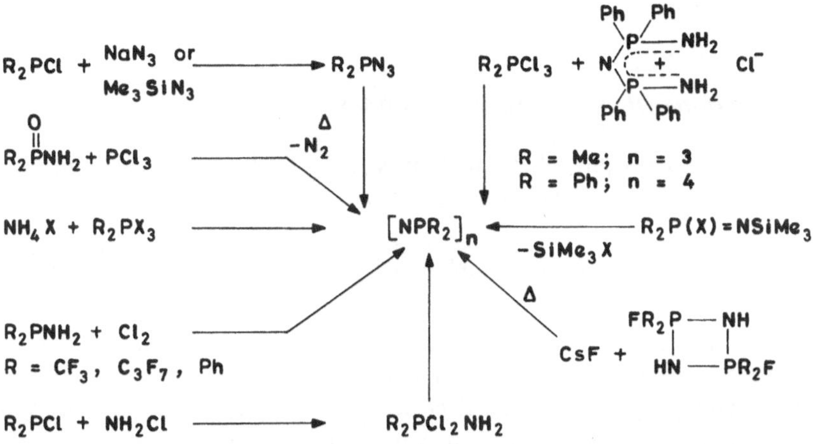

$$R_2PCl + \begin{array}{c} NaN_3 \ or \\ Me_3SiN_3 \end{array} \longrightarrow R_2PN_3$$

$$\underset{O}{\overset{\parallel}{R_2PNH_2}} + PCl_3 \xrightarrow[-N_2]{\Delta}$$

$$NH_4X + R_2PX_3 \longrightarrow$$

$$R_2PNH_2 + Cl_2$$
$$R = CF_3, C_3F_7, Ph$$

$$R_2PCl + NH_2Cl \longrightarrow R_2PCl_2NH_2$$

$$[NPR_2]_n$$

$$R = Me; n = 3$$
$$R = Ph; n = 4$$

$$R_2P(X) = NSiMe_3$$
$$-SiMe_3X$$

$$CsF + \begin{array}{c} FR_2P-NH \\ | \quad\quad | \\ HN-PR_2F \end{array}$$

Scheme 2

$$\underset{\parallel}{\overset{S}{Ph_2PNHCH_2Ph}} \xrightarrow{\quad\Delta\quad}$$

$$Ph_2P(Cl)NC(NH)NH_2 \cdot HCl \xrightarrow{\quad\Delta\quad} N_3P_3Ph_6 \tag{9}$$

$$[NH_2(PPh_2=N)_3H_2]^+Cl^- \qquad \Delta \qquad \uparrow$$

$$NH_4X + RPX_4 \longrightarrow N_3P_3R_3X_3 \tag{10}$$

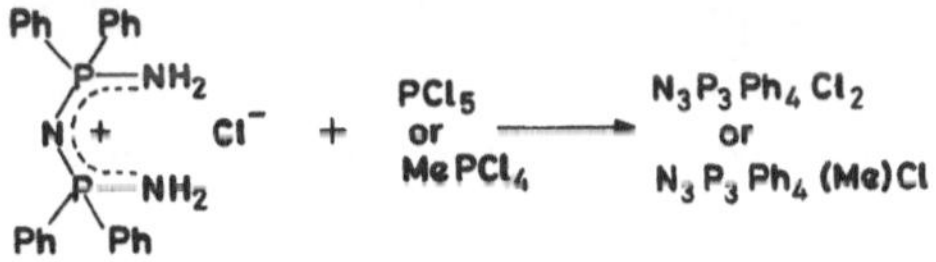

$$\begin{array}{c} + \ \begin{array}{c} PCl_5 \\ or \\ MePCl_4 \end{array} \longrightarrow \begin{array}{c} N_3P_3Ph_4Cl_2 \\ or \\ N_3P_3Ph_4(Me)Cl \end{array} \end{array} \tag{11}$$

Thermolysis of organo azido phosphines produce organo cyclophosphazenes in high yields [25]. Recently, Guy Bertrand and coworkers have obtained a four membered cyclophosphazene [26–27] derivative, $N_2P_2(NPr^i_2)_4$ by photolyzing the corresponding azidophosphine (Eq. 12).

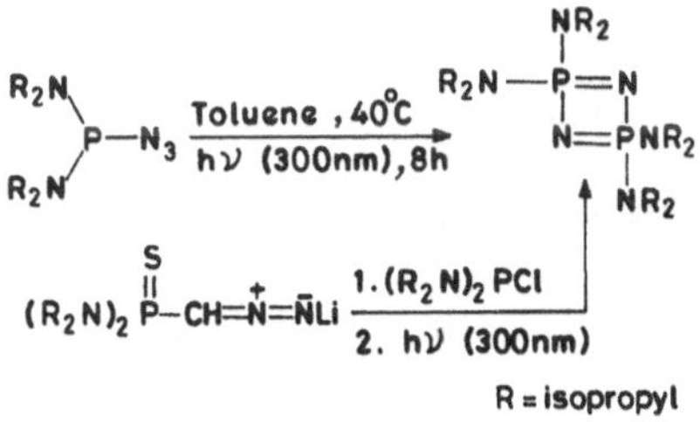

$$R = isopropyl \tag{12}$$

3 Reactions with Amines

3.1 Reactions of $N_3P_3Cl_6$ with Amines

The hexahalocyclotriphosphazenes $N_3P_3X_6$ (X = Cl, Br, F) can undergo substitution reactions with amines by cleavage of the P–X bond (Eq. 13) [3, 28]:

$$N_3P_3X_6 + RNH_2 \longrightarrow N_3P_3X_{6-n}(NHR)_n \qquad n = 1 \text{ to } 6 \qquad (13)$$

This reaction is quite general and reactions are known with ammonia, as well as with several primary and secondary amines [29–53]. However, the data on $N_3P_3F_6$ and $N_3P_3Br_6$ are scarce. There are probably two reasons for this. While in the case of $N_3P_3F_6$ the reactions with amines are very slow owing to the decreased lability of the P–F bond (vide infra), in the case of $N_3P_3Br_6$ the difficulty may be due to the tedious synthesis of the starting derivative.

The paucity of data with the fluoro and bromo derivatives is more than made up by the wealth of information available in the case of the chloro derivative $N_3P_3Cl_6$. It is likely that the general conclusions drawn from the study of the chloro derivatives can also be applied to the other halogeno derivatives. Some examples of the aminolysis reactions are illustrated in Table 1.

The general trends observed can be summarized as follows:

(a) Substitution reactions on $N_3P_3Cl_6$ with secondary amines afford mainly non-geminal products [28], the only exception being aziridine [29–33] where geminal and nongeminal products are formed in approximately equal amounts.

(b) The reactivity of $N_3P_3Cl_6$ towards primary amines is more complex. Ammonia itself reacts via a geminal pathway affording $2,2-N_3P_3Cl_4(NH_2)_2$ at the bis stage [34]. With other primary amines the nongeminal pathway is preferred [35–37]. However, increasing the steric bulk of the primary amine leads to a change in pathway and once again geminal products are formed [37, 48, 49]. Also, use of a tertiary amine increases the geminal products. There appears to be an electronic effect operating as well. Thus, while ethylamine affords nongeminal products, β-halo ethylamines give geminal products (in diethyl ether) [50, 51].

(c) Solvent effects also are extremely important and seem to play a crucial role in determining the stereo and regio-control [3, 52]. However, these effects are not understood. Also, amine hydrochlorides formed in the reaction assist in cis-trans isomerisations of the nongeminal products.

In one of the early crucial experiments towards unravelling the reaction pathways observed, Shaw has elegantly demonstrated that in the case of ethyl and tert-butylamine reactions with $N_3P_3Cl_6$ the incoming nucleophile determines the type of product formed [18, 53] (Eqs. 14–17)

Table 1. Reactions of selected amines (HNRR') with $N_3P_3Cl_6^{a,b}$

No.	Amine HNRR'		Products $N_3P_3(NRR')_nCl_{6-n}$			Ref.
	R	R'	n = 2	n = 3	n = 4	
1[c]	H	H	g	ni	ni	34
2[d]	H	Me	ng(t > c) > g	ni	ni	35, 36
3	H	Et	ng(t)	ng(tr)	g(tr)	37, 48
4	H	CH_2CH_2Cl	g	tr(u)	g	50
5	H	CH_2CH_2OMe	ng	tr(u)	g	50
6	H	CH_2CH_2COOEt	g	ni	g	51
7	H	CH_2Ph	ng(t) > g	ni	g	52b
8	H	Pr^i	ng(t > c) > g	ni	g	37, 48
9[e]	H	Ph	g, ng	s	g	38, 39
10[e,f]	H	$C_6H_4OCH_3$-p	ng(c = t), g	g = ng(c = t)	g	39
11	H	Bu^t	g	ni	g	49
12	Me	Me	ng(t > c) > g	ng(t > c) = g	ng(c)	41–43
13	Et	Et	ng(t > c)	ng(t > c) > g	ng(t > c)	40
14[f]	NC_2H_4		g > ng(c < t)	g > ng(c < t)	g > ng(c = t)	29, 30
15	NC_5H_{10}		ng(t > c)	ng(t) > g	ng(c)	44
16	Me	Ph	ng(c = t)	ng(c) = g	ni	45
17[d]	CH_2Ph	CH_2Ph	ng(u)	ni	ni	52b
18[d]	$N=PPh_3$		ng(u)	ni	ni	35b

[a] g = geminal, ng = nongeminal, t = *trans*, c = *cis*, tr = trace amount isolated, ni = not isolated or identified. s = spectroscopically identified in mixture but not isolated. u = isomers isolated, configuration unassigned.
[b] In most cases mono and persubstituted derivatives were isolated and pentakis derivatives remained elusive.
[c] mono derivative has been obtained indirectly by treating the geminal bis derivative with HCl in refluxing toluene.
[u] hexakis derivative was not isolated
[*] At the bis stage geminal product was obtained in THF with Et_3N as HCl scavenger.
[f] pentakis derivative $N_3P_3Cl(NRR')_5$ isolated.

$$N_3P_3Cl_5(NHEt) \xrightarrow[Bu^tNH_2 \cdot HCl]{+\ 2\,Bu^tNH_2} 2,2\text{-}N_3P_3Cl_4(NHEt)\,(t\text{-}BuNH) \qquad (14)$$

$$N_3P_3Cl_5(NHEt) \xrightarrow[-\ EtNH_2 \cdot HCl]{+\ 2\,EtNH_2} 2,4\text{-}N_3P_3Cl_4(NHEt)_2 \qquad (15)$$
$$(cis\ \&\ trans)$$

$$N_3P_3Cl_5(NHBu^t) \xrightarrow[-\ Bu^tNH_2 \cdot HCl]{Bu^tNH_2} 2,2\text{-}N_3P_3Cl_4(NHBu^t)_2 \qquad (16)$$

$$N_3P_3Cl_5(NHBu^t) \xrightarrow[-\ EtNH_2 \cdot HCl]{EtNH_2} 2,4\text{-}N_3P_3Cl_4(NHBu^t)(NHEt) \qquad (17)$$
$$(cis\ \&\ trans)$$

Thus *tert*-butylamine as an incoming nucleophile directs the formation of geminal products, while ethylamine favors nongeminal products.

Recently, detailed kinetic studies have been carried out on several systems which have unravelled the factors involved in the divergent reaction pathways observed [54–60]. There appear to be at least four types of mechanistic pathways involved. A proton abstraction mechanism (E_1CB) has been proposed to account for the geminal preference (Eq. 18).

$$(18)$$

In this mechanism, the formation of a three coordinate intermediate via the expulsion of HCl is proposed. Due to the enhanced electrophilicity of the tricoordinate phosphorus and with the need to relieve the strain, geminal products are preferred. Recently Krishnamurthy and coworkers have found kinetic evidence for this type of mechanism. The three coordinate intermediate has been trapped and its derivatives isolated [59].

In general, the formation of the non-geminal products at the bis stage would be anticipated from steric as well as electronic grounds. Thus, the presence of a –NHR group on P(2) would deactivate P(2) towards further nucleophilic substitution, due to the electron release by the exocyclic group. Also, the presence of an amino substituent on P(2) would sterically crowd the transition state, leading to the eventual product. At least three types of transition states are recognized. The $S_N(2)$ mechanism can operate in two ways:

(a) Formation of a neutral five coordinate phosphorus intermediate followed by the expulsion of the leaving group.
(b) A concerted $S_N(2)$, akin to the mechanism operating in carbon systems. Here the transition state is clearly polar.

The $S_N(2)$ reactions are characterized by a $\Delta H^{\#}$ term associated with the ease of formation of the neutral five coordinate intermedite, and a $\Delta S^{\#}$ term which should be invariant if everything else is constant. However, changing the reaction medium from a non-polar to a polar solvent would result in a favorable $\Delta S^{\#}$ owing to solvation of the polar transition state. Krishnamurthy and coworkers have found a gradual mechanistic change in the reactions of $N_3P_3Cl_6$ with dimethylamine (Scheme 3). Increase of substituents would be expected to increase $\Delta H^{\#}$ in non-polar solvents. However, in polar solvents where concerted mechanisms are dominant, favorable $\Delta S^{\#}$ directs the reaction pathway. In the conversion of 2,4,6-trans-$N_3P_3Cl_3(NMe_2)_3$ to 2,2,4,6-cis $N_3P_3Cl_2(NMe_2)_4$ the steric crowding at all the phosphorus centers combined with the total electron release of the –NMe_2 substituents causes a switch over in the mechanism and a dissociative S_N1 pathway is preferred (Eq. 19). This effect is also realized in the conversion of $N_3P_3(OPh)_5Cl$ to $N_3P_3(OPh)_5(NMe_2)$ (Eq. 20).

$$N_3P_3Cl_3(NMe_2)_3 + HNMe_2 \longrightarrow 2,2,4\text{-}trans\text{-}6\text{-}N_3P_3Cl_2(NMe_2)_4 \quad (19)$$

$$N_3P_3(OPh)_5Cl \xrightarrow[2\,HNMe_2]{} N_3P_3(OPh)_5(NMe_2) \quad (20)$$

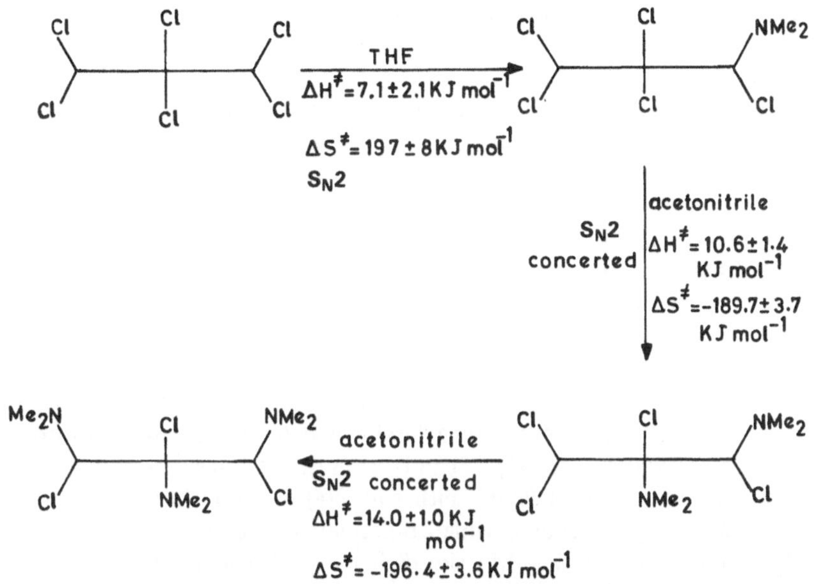

THF
$\Delta H^{\ddagger} = 7.1 \pm 2.1$ KJ mol^{-1}
$\Delta S^{\ddagger} = 197 \pm 8$ KJ mol^{-1}
S$_N$2

S$_N$2 concerted
acetonitrile
$\Delta H^{\ddagger} = 10.6 \pm 1.4$ KJ mol^{-1}
$\Delta S^{\ddagger} = -189.7 \pm 3.7$ KJ mol^{-1}

acetonitrile
S$_N$2 concerted
$\Delta H^{\ddagger} = 14.0 \pm 1.0$ KJ mol^{-1}
$\Delta S^{\ddagger} = -196.4 \pm 3.6$ KJ mol^{-1}

Scheme 3.

Here also, the combined electron release of the five phenoxy substituents along with the steric constraints of the substrate $N_3P_3(OPh)_5Cl$ allow the reaction to proceed by a S_N1 pathway. The increased amounts of geminal isomers at higher substitution stages with primary amines is probably due to the predominance of the S_N1(CB) pathway.

The decreased reactivity of $N_3P_3F_6$ in comparison with $N_3P_3Cl_6$ has now been rationalized as due to a strong $p\pi$-$d\pi$ bonding interaction between the exocyclic substituent and phosphorus. This results in a rigid structure for the parent molecule. Since a structural reorganization is required in $S_N(2)$ mechanisms, a slower reactivity is observed. In the case of aziridine where the exocyclic substituent is small and also weakly electron releasing both $S_N(1)$ and $S_N(2)$ pathways might be operating leading to the observed formation of geminal as well as non-geminal products [29–33]. For a more exhaustive survey of the reaction pathways in cyclophosphazenes the reader is referred to recent review [1].

3.2 Reactions of $N_4P_4Cl_8$ with Amines and Formation of Bicyclic Phosphazenes

$N_4P_4Cl_8$ reacts much faster than $N_3P_3Cl_6$ [61–67]; for example in the reaction with *tert*-butylamine $N_4P_4Cl_8$ has been shown to react at least 200 times faster than $N_3P_3Cl_6$ [57]. The main reason for this enhanced reactivity appears to be the greater skeletal flexibility of the eight membered ring. Although detailed kinetic studies are not available it appears that the reaction of amines with $N_4P_4Cl_8$ proceeds by a bimolecular mechanism. The skeletal flexibility of $N_4P_4Cl_8$ allows the formation of the neutral five coordinate intermediate readily thereby lowering the enthalpy of activation for this process. This leads to a predominantly nongeminal reaction pathway. The greater reactivity of $N_4P_4Cl_8$ is also manifested in its reactions with bulky amines. Thus, for example, $N_3P_3Cl_6$ reacts with dibenzylamine even under harsh conditions to afford substitution of only two chlorines. In contrast, $N_4P_4Cl_8$ reacts with dibenzylamine more readily and substitution up to four chlorines are realized. [67].

While reactive amines such as dimethylamine, ethylamine, etc. afford predominantly 2-*trans*-6 derivatives at the bis stage [62, 64, 65], sluggishly reacting amines such as *tert*-butylamine, benzylamine and *N*-methyl aniline afford both 2,4- and 2,6-derivatives [61–63]. It can be seen that 2,4-derivatives are likely to be formed by an attack of RNH_2 either at P(4) or P(8) (thereby making this process statistically favorable), while 2,6-derivatives are formed by an attack of RNH_2 preferentially at the P(8) site (Scheme 4). Strongly electron releasing groups such as Me_2N- or Et_2N- would deactivate the P(4) and P(8) sites leaving the P(6) site as the preferred position for substitution. The reactions of $N_4P_4Cl_8$ with some amines are summarized in Table 2. It should be noted that

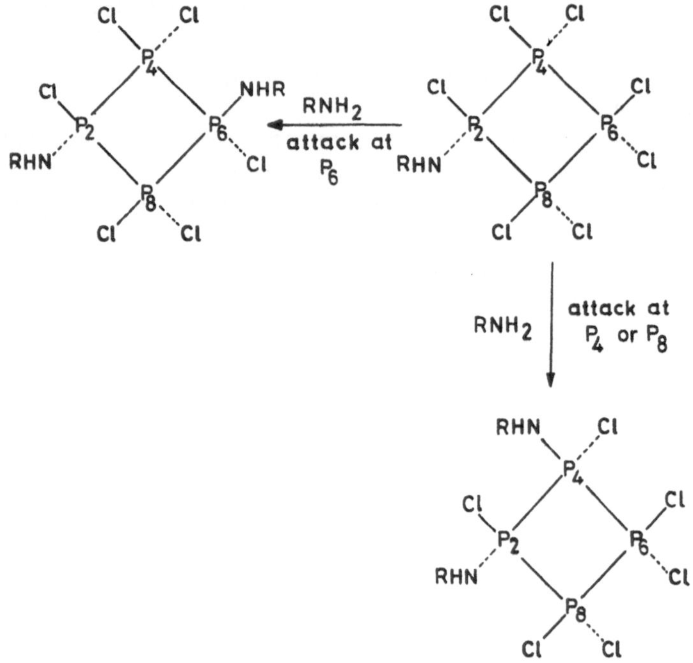

Scheme 4

Table 2. Reactions of amines HNR R' with $N_4P_4Cl_8$[a]

No	Amine HNRR'		Products $N_4P_4Cl_{8-n}(NRR')_n$			
	R	R'	n = 2	n = 3	n = 4	n = 5
1[b]	H	Et	2, t-6	2, c-4, t-6	2, c-4, t-6, t-8	ni
					2, c-4, c-6, t-8	
2	H	Bu^t	2, 6 & 2, 4(u)	2, 4, 6(u)	ni	ni
3[d]	Me	Me	2, t-6	2, c-4, t-6	2, t-4, t-6, t-8	2, 2, 4, t-6, t-8
				2, t-4, t-6	2, c-4, t-6, c-8	2, 2, 4, 6, 6
				2, 2, 6	2, c-4, t-6, t-8	
					2, 2, 6, 6	
4[d, e]	Me	Ph	2, t-6	2, c-4, c-6	2, c-4, t-6, t-8	ni
			2, t-4	2, t-4, t-6	2, 2, 6, 6	
					2, c-4, c-6, t-8	
					2, t-4, t-6, t-8	
					2, 2, 4, t-6	
5[e, b]	CH$_2$Ph	CH$_2$Ph	2, t-6	ng(u)	ng(u)	ni
			2, t-4			
6[e]	NC$_2$H$_4$		2, 2 <	2, 2, 6	ni	ni
			2, 4 (c > t) <	2, t-4, c-6		
			2, 6 (t > c)	2, c-4, t-6		
				2, 2, 4^f		
				2, c-4, c-6^f		

[a] In most cases mono and octakis derivatives are isolated. In all reactions heptakis derivative remained elusive. Legend: c = cis, t = trans, ni = not isolated, u = isomers obtained but structures unassigned.
[b] Bicyclic derivative of formula $N_4P_4(NRR')_6(NR')$ isolated.
[d] Hexakis derivative of configuration 2, 2, 4, t-8, 6, 6 isolated.
[e] Octakis derivative not isolated.

of the possible 33 isomers ten have been isolated for N-methyl aniline. A unique feature of this study is the isolation of five isomers out of a possible ten at the tetrakis stage [63].

In addition to the 'normal' pathways available in the reactions of eight membered $N_4P_4Cl_8$ with amines, a new reaction mode to afford bicyclic phosphazenes has been found. The reaction of 2,6-trans-$N_4P_4Cl_6(NHEt)_2$ with dimethylamine in ether proceeds to afford the expected 'normal' product 2,6-$N_4P_4(NMe_2)_6(NHEt)_2$. However, the reaction proceeds in an entirely different manner in chloroform and affords substantial yields of a new derivative, a bicyclic phosphazene, in which the P(2) and P(6) positions are bridged by an amino group [68] (Eq. 21).

(21)

This reaction pathway has been shown to be completely general and occurs in the reactions of any 2,6-primary amino cyclotetraphosphazene derivative with primary or secondary amines in polar solvents such as chloroform and acetonitrile [69–71]. It has been shown, that the bicyclic product is formed only after the tetrakis stage of substitution in the cyclotetraphosphazenes. Thus 2,6-$N_4P_4Cl_6(NHEt)_2$ reacts with $HNMe_2$ to give 'normal' 2,6,4,8-$N_4P_4Cl_4(NHEt)_2(NMe_2)_2$. This derivative on further reaction with $HNMe_2$ in chloroform gives the bicyclic product (Scheme 5) [71]. The mechanism of the bicyclic product formation has been postulated to occur via a single or a double proton abstraction mechanism (Scheme 6) [68]. Polar solvents would stabilize the species formed after proton abstraction by hydrogen bonding to the nitrogen involved in intramolecular attack, thereby explaining formation of bicyclic derivatives in polar solvents. Further, the yields of the bicyclic product increases in the solvents $CH_3CN \ll CH_2Cl_2 < CHCl_3$ which is in the same order as the acidic nature of the proton. It is expected that the acidic protons facilitate a heterolysis of the P–Cl bond by solvation of $\equiv PCl(NHR)$. Further, in the reactions of 2,6-$N_4P_4Cl_6(NHR)_2$ with $HNMe_2$ to afford bicyclic products $N_4P_4(NMe_2)_5(NHR)(NR)$ the yields decrease with increased steric bulk. Thus,

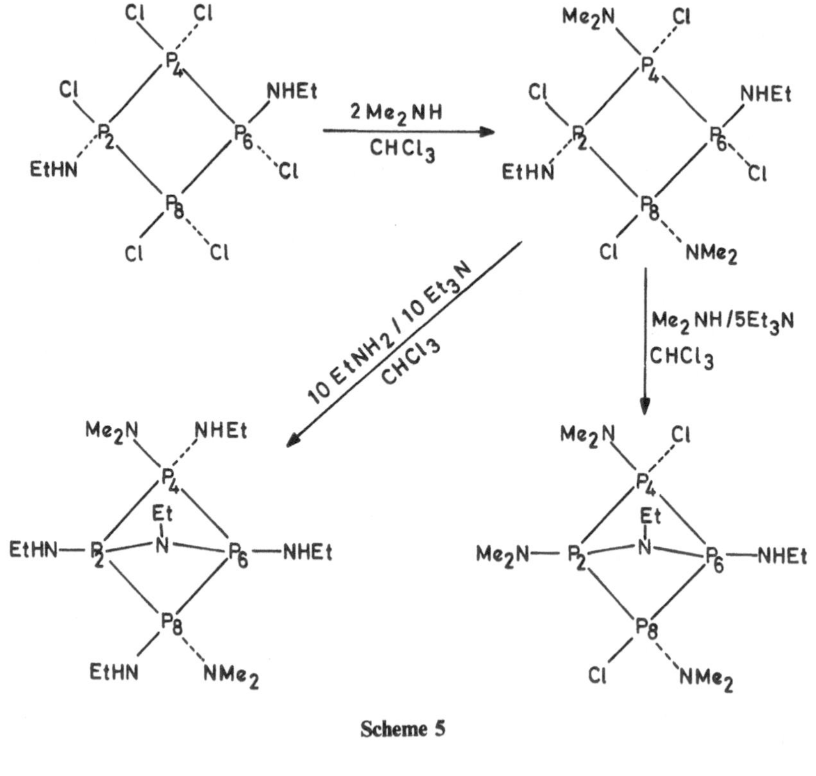

Scheme 5

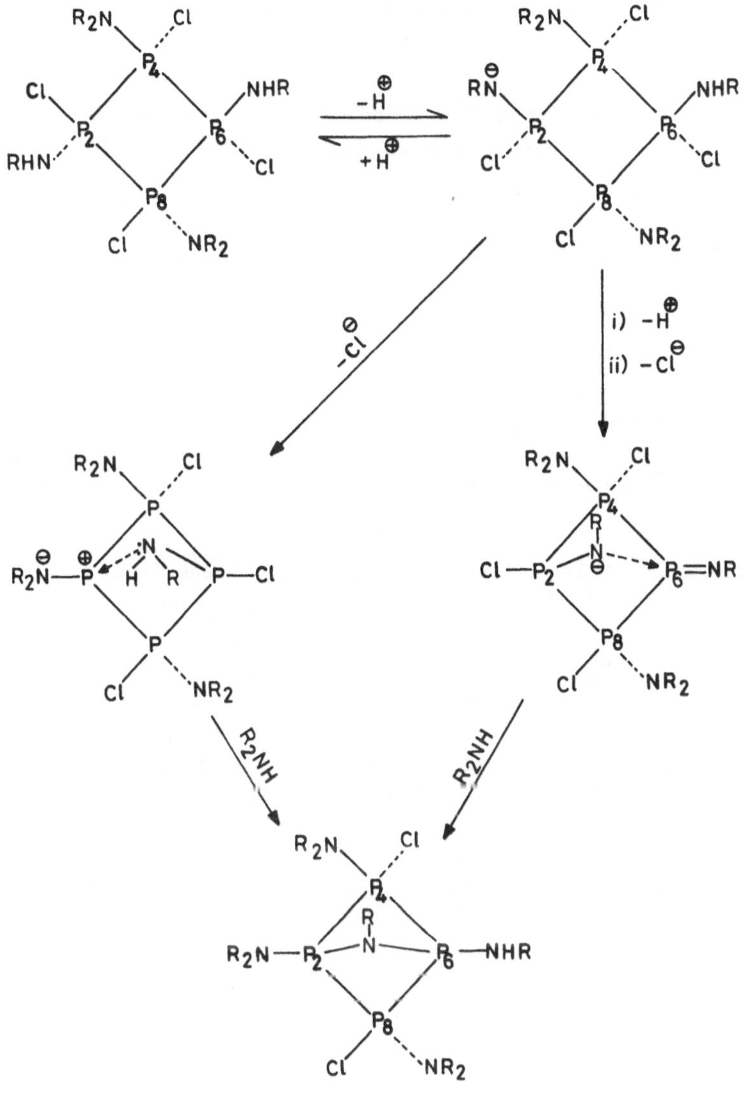

Scheme 6

with $R = Bu^n$, yield is low, while when $R = Bu^t$ or Pr^i, no bicyclic formation takes place [69].

Another route to the bicyclic phosphazenes is through a dealkylation-cum-transannular attack; in the reactions of $N_4P_4Cl_8$ with the bulky dibenzylamine biyclic phosphazenes are formed where the bridgehead is an amino substituent which has underwent a dealkylation [67].

4 Reactions with Hydroxy Compounds and Mercaptans

Alcohols and phenols react with halocyclophosphazenes in the presence of a base to give alkoxy/aryloxy cyclophosphazenes (Eq. 22) [1–3].

$$N_3P_3Cl_6 + 6\,RONa \longrightarrow N_3P_3(OR)_6 + NaCl \tag{22}$$

Fully substituted derivatives are easily obtained by treating excess of alcohol or phenol with the halocyclophosphazene [72–79]. However, the iosolation of partially substituted derivatives remains very difficult. Allcock and coworkers isolated a series of trifluoroethoxy cyclotriphosphazenes by GLC technique (Eq. 23) [73a]. Alcohols and phenols react in a nongeminal pathway with halocyclophosphazenes [3].

$$N_3P_3Cl_6 + nCH_3CF_2ONa \xrightarrow{\ n=1\ \text{to}\ 6\ } N_3P_3Cl_{6-n}(CH_3CF_2O)_n$$

$$\tag{23}$$

Recently, Allcock and coworkers have observed a geminal pathway operating in the reactions of 2,6-dichloro phenol with $N_3P_3Cl_6$ [73b]. They have invoked an S_N1 mechanism to explain the formation of geminal products. The preference of nongeminal pathway with 2,6-dimethyl phenol indicates that the steric bulk is not the factor which dictates the mechanism [73b]. However, it is difficult to ascertain a particular mechanism with the scarce literature evidence. More elaborate studies are required to unravel the pathways involved in the reactions of alcohols or phenols with halocyclophosphazenes.

Reactions of thiolates with chlorocyclophosphazenes are found to proceed in a geminal pathway. Only very limited thiols such as thiophenol, methyl and ethyl mercaptans have been subjected to these studies (Eq. 24) [80, 81].

$$N_3P_3Cl_6 + nRSNa \longrightarrow N_3P_3Cl_{6-n}(SR)_n; \quad n = 2, 4, 6 \tag{24}$$

$$R = Me, Et, Ph$$

5 Reactions with Difunctional Reagents

The reactions of several difunctional reagents such as diamines, diols and amino alcohols with chlorocyclophosphazenes in general and $N_3P_3Cl_6$ in particular have been studied in great detail [82–84]. While $N_4P_4Cl_8$ is much more reactive than $N_3P_3Cl_6$ substituted derivatives such as 2,2-$N_3P_3Cl_4Ph_2$ and 2,2-$N_3P_3Cl_4(NHBu^i)_2$ react more slowly [85]. In all these reactions there appear to be four distinct reaction pathways (Scheme 7), leading to spirocyclic, *ansa*, intermolecular bridged or open chain products.

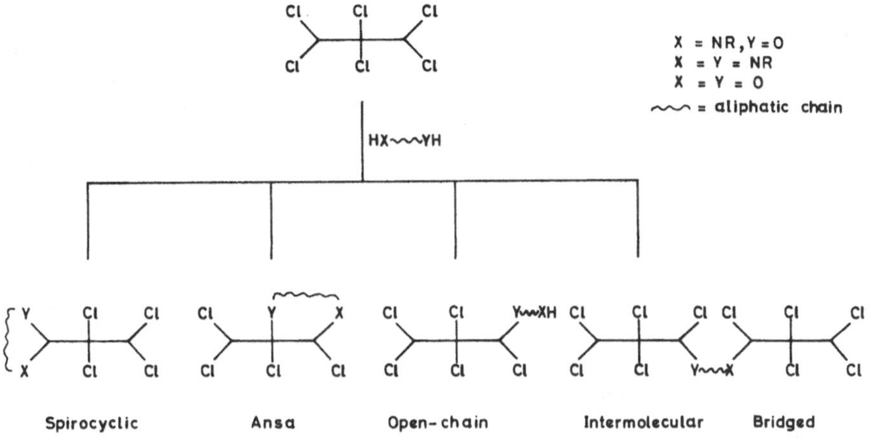

X = NR, Y = O
X = Y = NR
X = Y = O
∿∿ = aliphatic chain

Spirocyclic   Ansa   Open-chain   Intermolecular   Bridged

Scheme 7

Spirocyclic product is formed as a result of replacement of both the chlorines on the same phosphorus atom, while the *ansa* product is formed as a result of the replacement of chlorines from two distinct phosphorus atoms within the same molecule. In open chain product only one end of the difunctional reagent is involved in reaction with the chlorocyclophosphazene. Finally, intermolecular bridged products result from the reaction of two chlorocyclophosphazene molecules with a difunctional reagent. This last reaction also is a model reaction for condensation polymerization involving phosphazenes.

Trimeric chloride and its derivatives 2,2-$N_3P_3Cl_4Ph_2$ and 2,2-$N_3P_3Cl_4(NHBu^t)_2$ react with 1,2-diamino ethane, 1,3-diamino propane 1,4-diamino butane, 1,2-ethane diol, 1,3-propane diol, ethanolamine, propanolamine to afford predominantly spirocyclic products [85]. The stability of the spirocyclic product formed seems to be governed by (a) spirocyclic ring size (five, six and seven membered rings are more stable) (b) presence of reactive residual NH's and chlorines in the product. This latter feature can lead to cross linking reactions. Removal of either of the reactive groups can render greater product stability (Scheme 8). The competing crosslinking reaction can also inhibit

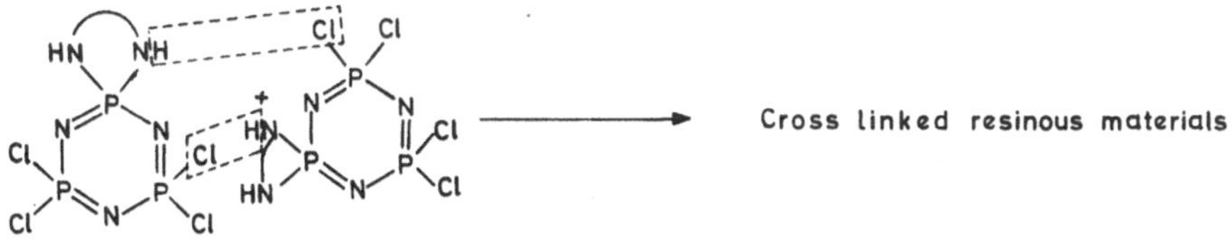

Cross linked resinous materials

Scheme 8

formation of di and trispiro derivatives. Thus, for example, while 1,2-diamino ethane reacts with $N_3P_3Cl_6$ to give only monospiro derivative [86], N,N'-dimethyl-1,2-diamino ethane affords all the three stages of substitution and mono, di and trispiro derivatives are isolated (Scheme 9) [87, 88]. In the case of unsymmetric difunctional reagents such as N'-methyl ethanolamine both *cis* and *trans* isomers have been characterised at the di and trispiro stages (Fig. 3) [89, 90]. Ansa and open chain products have remained rare. Harris and cowor-kers have isolated both these examples in the reaction of $N_3P_3Cl_6$ with 1-amino-2-propanol (Eq. 25) [91]. Shaw and coworkers have found that in the reactions of $N_3P_3Cl_6$ with diols although predominantly spirocyclic products result, traces of *ansa*, open chain and intermolecular bridged derivatives are also formed [92–100]. In the reaction of $N_3P_3Cl_6$ with 1,3-propane diol an inter-esting spiro-*ansa* product has been isolated at the four chlorine replacement stage [98]. In a series of studies with long chain aliphatic diamines and oxodiamines with $N_3P_3Cl_6$, Labarre and coworkers have been able to isolate spirocyclic, *ansa* as well as intermolecular bridged products [84, 101–130]. They have found an interesting solvent dependence on the reaction pathway (Scheme 10). However, the reasons for this solvent dependence are not clear.

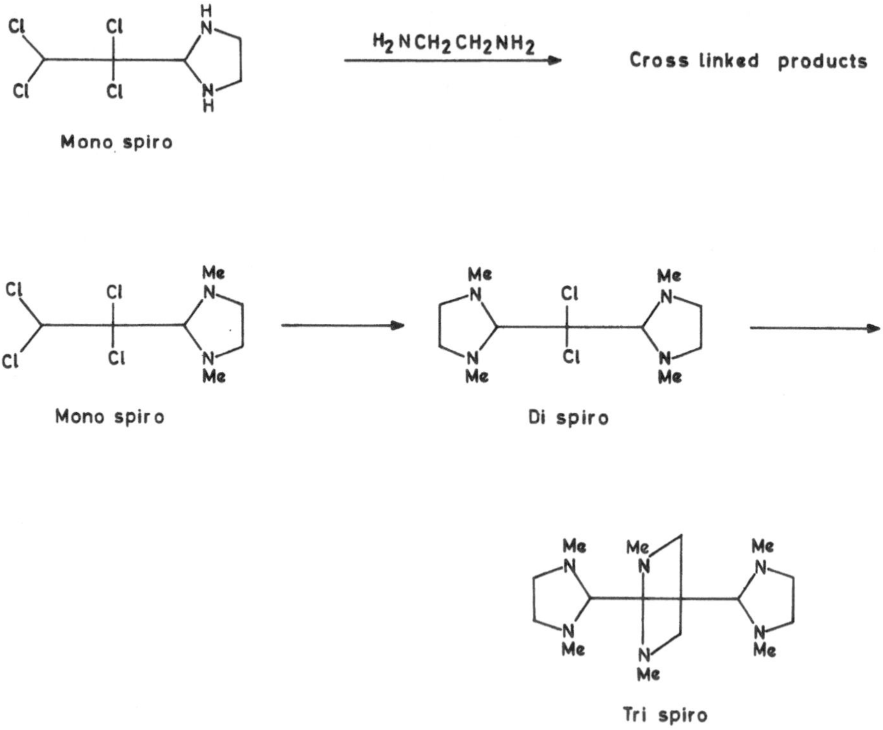

Scheme 9

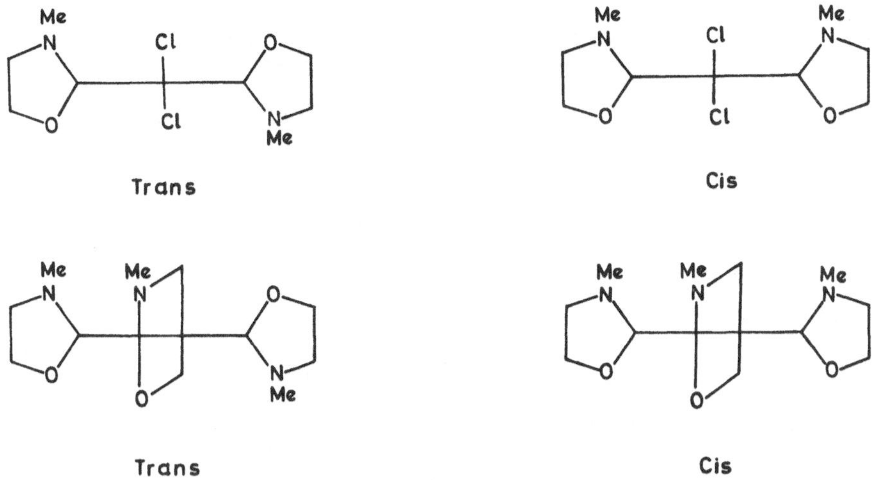

Fig. 3. Formation of *cis* and *trans* isomers for di and tri spirocyclic products $N_3P_3Cl_2$ $[N(Me)CH_2CH_2O]_2$ and $N_3P_3[N(Me)CH_2CH_2O]_3$

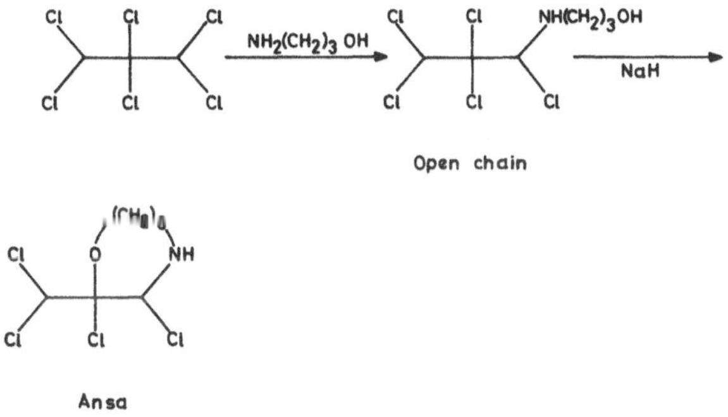

(25)

Several aromatic difunctional reagents react with $N_3P_3Cl_6$ to give trispiro derivatives [131–134]. Many of these decompose readily to cyclic phosphoranes. Recently attempts at restricting substitution to the mono and dispiro derivatives have been successful in the reactions of α,α'-dihydroxy binaphthyl [135]. Allcock and coworkers have shown that the tris catecholate derivative $N_3P_3(O_2C_6H_4)_3$ forms inclusion clathrates with several small molecules [136]. Certain acrylate monomers can also be trapped in the channels of the tris catecholate derivative, and can be polymerised in a stereoregular manner [137]. Clathrate formation has also been reported to occur with $N_3P_3(Az)_6 (Az = 1$-aziridinyl) [1].

In contrast to $N_3P_3Cl_6$, $N_4P_4Cl_8$ is extremely reactive towards difunctional reagents. This has led to the isolation of several decomposition products. Reactions with N-methyl ethanolamine [87], 1,3-propane diol and 1,3-diamino propane afford mainly spiro products [138]. A detailed investigation on the reactions of $N_4P_4Cl_8$ with $HO-(CH_2)_n-OH$ (n = 3, 4) has revealed that

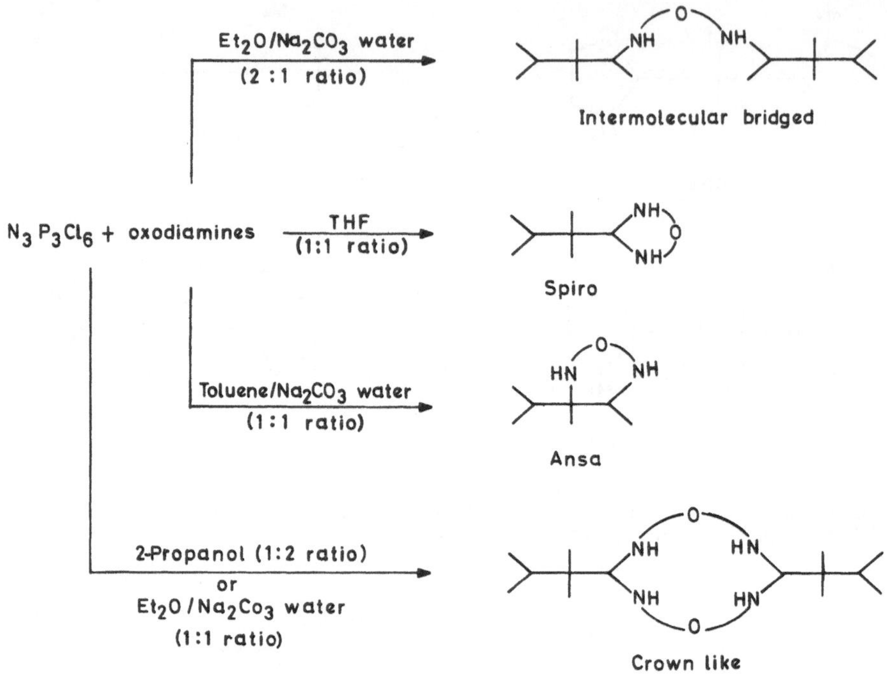

Scheme 10

spirocyclic products (mono, di, tri and tetra) are formed at all stages of chlorine substitution [95, 139, 140].

6 Isomerization, Tautomerism and Thermal Rearrangements

6.1 cis ↔ trans Isomerisation

Non-geminal bis, tris, and tetrakis secondary amino derivatives are found to undergo reversible *cis* ↔ *trans* isomerisations in the presence of amine hydrochlorides [141–145]. It is reported that aluminium trichloride also acts as a catalyst for these interconversions (Eq. 26) [142].

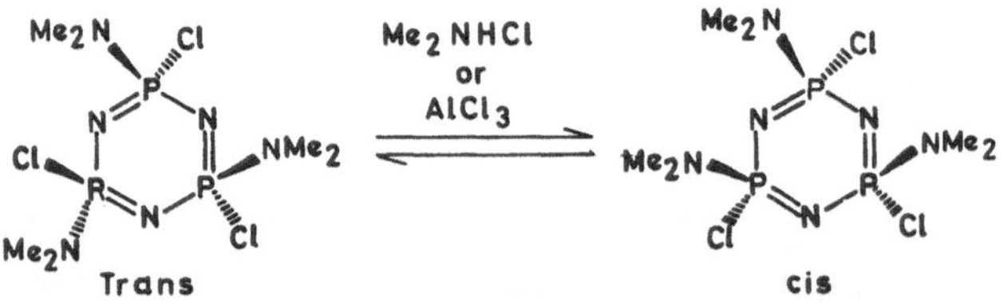

$$\tag{26}$$

Allcock and coworkers have observed similar *cis* ↔ *trans* transformations in the reactions of Grignard reagents with the amino cyclophosphazenes. Reactions of *cis*-$N_3P_3Cl_2(NMe_2)_4$ with $RMgBr(R = Me$ or $Et)$ in the presence of Et_3N gives a 50:50 mixture of *cis*- and *trans*-$N_3P_3R_2(NMe_2)_4$ (Eq. 27) [146]. The role of amine hydrochlorides or $AlCl_3$ in the *cis* ↔ *trans* isomerism is not properly understood and more elaborate experiments are needed to delineate the mechanism fully [145].

$$2,4\text{-}N_3P_3Cl_2(NMe_2)_4 \xrightarrow[\text{Et}_2\text{O}]{\text{RMgBr/Et}_3\text{N}} 2,4\text{-}N_3P_3R_2(NMe_2)_4 \qquad (27)$$

'cis' 'cis & trans'

6.2 Tautomerisations (O ↔ N Hydrogen Migrations)

Hydrolysis of chlorocyclophosphazenes affords hydroxy oxophosphazanes via a rapid tautomerism of the initially formed dihydroxy phosphazenes (Eq. 28) [147–149]. Crystal structure investigations on $N_3P_3(OPh)_5(O)(H)$ [150],

$$N_3P_3Cl_6 \xrightarrow{\text{H}_2\text{O}} N_3P_3(OH)_6 \xrightarrow{\text{tautomerism}} N_3P_3O_3(OH)_3H_3 \qquad (28)$$

'hydroxy 'hyroxy *oxo*
phosphazene' phosphazane'

$N_3P_3Ph_2(OMe)_3(O)(H)$ [151] and $N_3P_3(NEt_2)Cl_2(O)(H)$ [152], reveal that all of these have the cyclophosphazadiene structures:

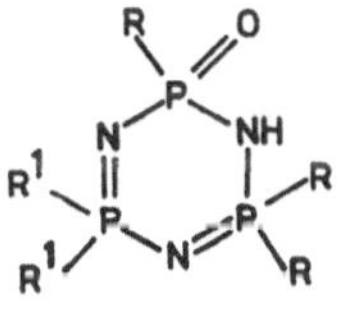

R = OMe ; R^1 = Ph
R = R^1 = OPh
R = NEt$_2$; R^1 = Cl

In solution it has been possible to identify different tautomeric structures [151]. For example, the *gem*-$N_3P_3Ph_2(OR)_3(O)(H)$ $(R = Me$ or $Pr^n)$ has been shown to be present in two tautomeric forms while *gem*-$N_3P_3(NHBu^i)_2(OMe)_3(O)(H)$ has only one preferred tautomeric form.

In principle a mono derivative, *gem*-$N_3P_3R_2R_3'OH$ should have four tautomeric forms (Fig. 4). Apart from the trivial hydroxy form (I), two tautomers (II and III) arise from the movement of hydrogen to the two non-equivalent α-ring nitrogens. The fourth tautomer (IV) is due to the migration of hydrogen to the γ-ring nitrogen atom with reference to the phosphoryl group.

The proton migration to the α-ring nitrogen has been investigated by variable temperature ^{31}P NMR spectroscopy [151, 153]. Three types of tautomerism involving α-ring nitrogen atoms have been experimentally observed.

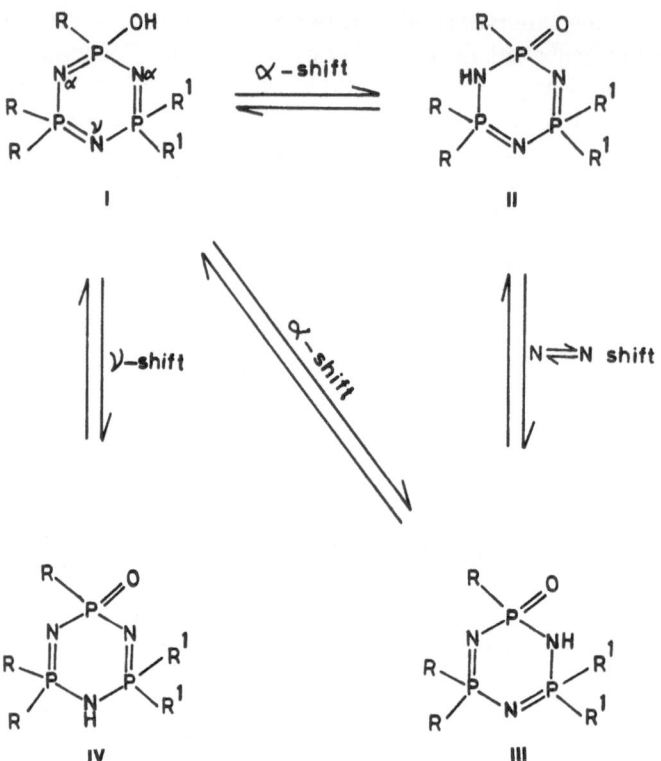

Fig. 4. Tautomers derived from monohydroxy cyclotriphosphazene, $N_3P_3R_3R'_2(OH)$

These are related to the relative ring nitrogen basicities:

(a) when a particular ring nitrogen site is more basic than the other, no N↔N hydrogen shift is observed and the proton is preferentially attached to the most basic site. The ^{31}P NMR spectrum for these derivatives is of the spin system ABX and temperature invariable. An example of this type is *gem*-$N_3P_3(NHBu^t)_2(OMe)_3(O)(H)$; the proton migrates to the α-ring nitrogen adjacent to the phosphorus containing *tert*-butylamino groups (Fig. 5) [153].

(b) Two equivalent α-ring nitrogen sites [example: $N_3P_3(OPh)_5(O)(H)$] cause rapid N↔N hydrogen exchange thus giving an A_2X spectrum. Arresting the exchange by lowering the temperature results in an ABX pattern in the ^{31}P NMR spectrum (Fig. 6) [153].

(c) Hydrogen transfer to two non-equivalent α-ring nitrogen sites would result in two ABX spectra in their ^{31}P NMR indicative of two isomers in solution (Fig. 4: II and III; R = R'), when there is no exchange of hydrogen between them. This is observed for *gem*-$N_3P_3Ph_2(OPr^n)_3(O)(H)$ at $-40\,°C$. Fast

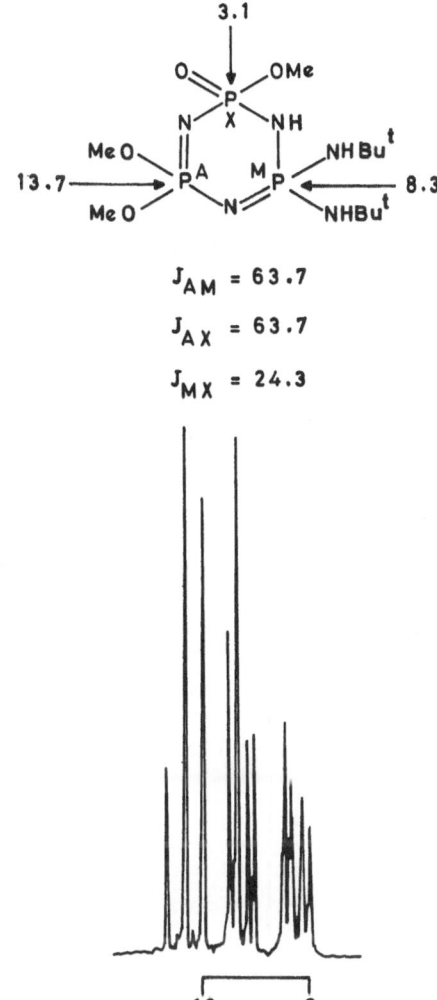

Fig. 5. ^{31}P NMR spectrum of *gem*-N$_3$P$_3$ (NHBu^t)$_2$(OMe)$_3$(OH) (from Ref. 153)

exchange at 100 °C results in a single ABX pattern and at ambient temperatures where intermediate rates of exchange are expected a broad unresolved pattern is observed in the ^{31}P NMR [153].

The γ-form(IV) has so far remained elusive. The rate of tautomerisation is proposed as playing a key role in determining the hydrolysis pathway. Thus, if the rate of tautomerisation is slower than the subsequent hydrolysis, nucleophilic attack at the same phosphorus leads to a geminal dihydroxy derivative. This undergoes tautomerisation to give a hydroxy oxophosphazadiene. Alternatively faster tautomerisation leads to nongeminal attack (Scheme 11) [154].

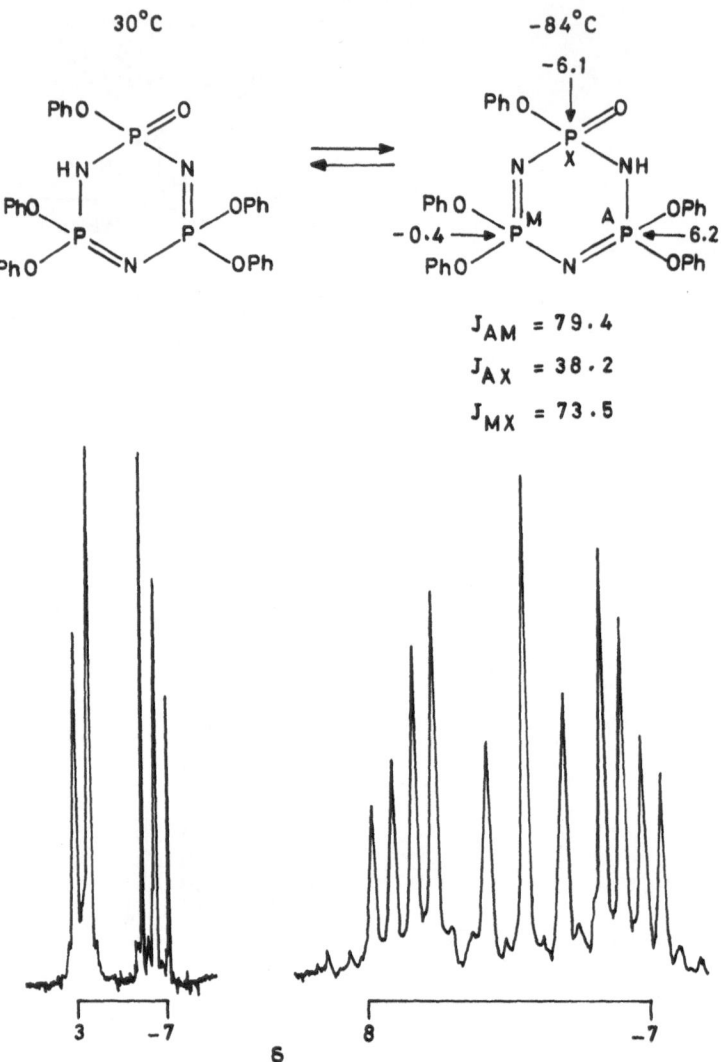

Fig. 6. ^{31}P NMR spectra of $N_3P_3(OPh)_5(OH)$ at 30 °C and − 84 °C (from Ref. 153)

6.3 Rearrangements of Alkoxy Cyclophosphazenes (O↔N Alkyl Migrations)

Alkoxy cyclophosphazenes undergo rearrangements to the corresponding oxy-cyclophosphazanes by an irreversible migration of an alkyl group from oxygen to the ring nitrogen [155–164]. The following is a summary of the available

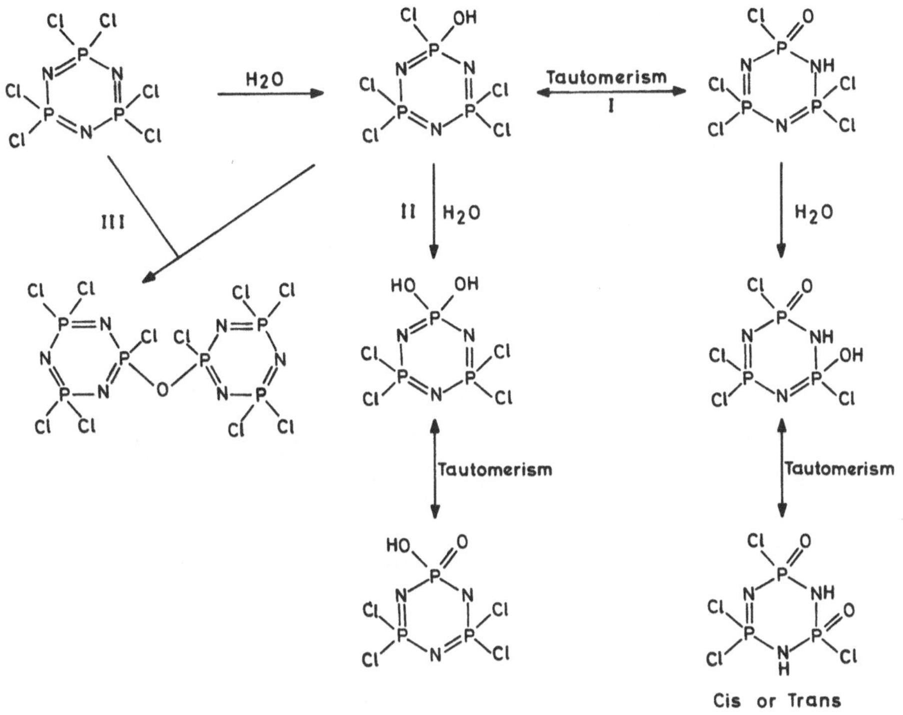

Scheme 11

experimental evidence:
(a) In contrast to the simple tautomerism discussed earlier alkyl shifts occur to the γ-ring nitrogen also.
(b) Fully substituted methoxy cyclophosphazenes [NP(OMe)$_2$]$_n$ and partially methoxylated nongeminal derivatives [for example, N$_3$P$_3$(NMe$_2$)$_2$(OMe)$_4$] give only the fully rearranged products viz., trioxaphosphazanes [161]. In contrast, the geminal methoxy derivatives such as N$_3$P$_3$R$_2$(OMe)$_4$ (R = Ph or NHBu^t) yield both fully and partially rearranged products, the dioxaphosphaz-1-enes and oxophosphazadienes (Scheme 12) [161].
    This can be correlated with the relative basicity of the ring nitrogens.
(c) The geometrical disposition of the substituents are unaltered upon thermal rearrangement. Thus for example the *cis* and *trans* isomers of 2,4,6,2,4,6-N$_3$P$_3$(OC$_6$H$_4$Me-*p*)$_3$(OMe)$_3$ undergo thermolysis with retention of configurations [165].
(d) These rearrangements are generally sluggish at the beginning of the reactions and can be initiated by the addition of alkyl halides or rearranged *oxo* products [164]. These results have been rationalized by suggesting that rearrangement in the initial stage proceeds by a slow intermolecular pathway whilst the subsequent steps involve a relatively faster intramolecular mechanism. Existence of intermolecular mechanism is supported by the

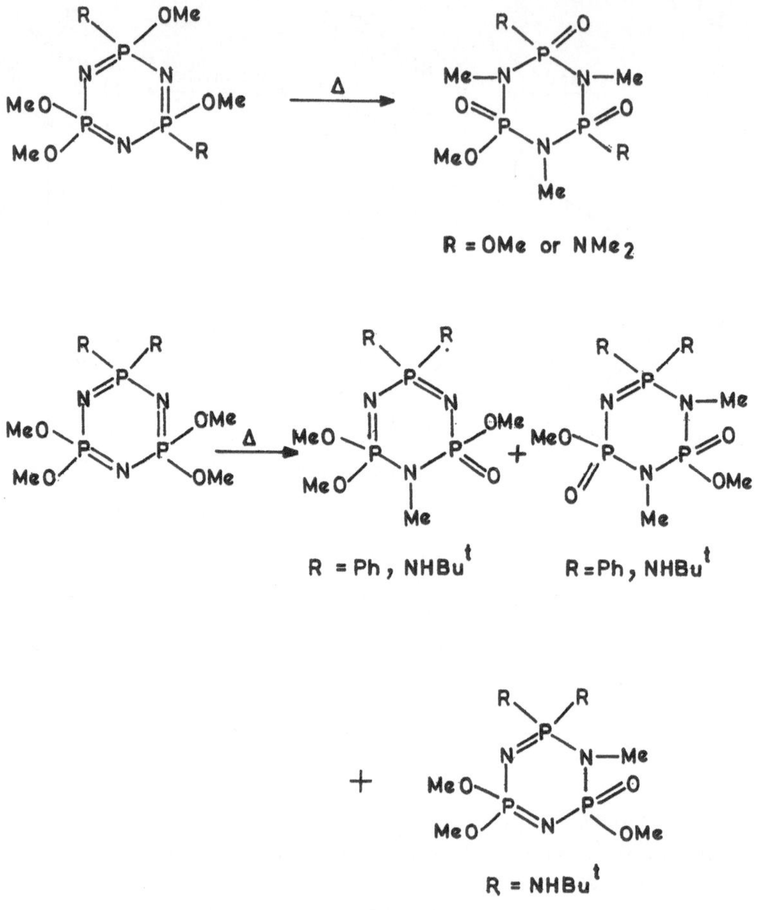

R = OMe or NMe$_2$

R = Ph, NHBu^t R = Ph, NHBu^t

R = NHBu^t

Scheme 12

observation that scrambled products are identified when $N_3P_3(OMe)_6$ and $N_3P_3(OCD_3)_6$ are subjected together to thermolysis [166]. Similarly, on heating nongem trans-$N_3P_3(OCH_2Ph)_3(OC_6H_4Me$-$p)_3$ along with nongem trans-$N_3P_3(OMe)_3(OC_6H_4Me$-$p)_3$ scrambled products are isolated in addition to the unscrambled products [167].

(e) Fluoro alkoxy and aryloxy cyclophosphazenes which are highly flame retardant are resistant to thermal rearrangements.

6.4 Other Miscellaneous Rearrangements

The gem-$N_3P_3(NH_2)_2Cl_4$ when allowed to react with alkoxide ion in the corresponding alcohol gives rearranged cis and trans alkoxy derivatives, nongem-$N_3P_3(NH_2)_2(OR)_4$ (R = Me, Et, Pr^n or Bu^n) (Eq. 29) [168–170]. However, this rearrangement is not observed in the presence of a tertiary amine base. The mechanism of this remarkable amino group migration is not conclusively understood. However, the formation of *cis* and *trans* isomers points to an intermolecular process.

$$ (29) $$

$$ R = Me, Et, Pr^n \, \& \, Bu^n $$

Another interesting rearrangement has been reported by Paddock and coworkers [171, 172]. The reaction of sodium bis(trimethylsilyl)amide with *N*-methyl phosphazenium iodides $N_nP_nMe_{2n+1}I$ (n = 3, 4) gives the corresponding phosphorin derivatives (Eq. 30) via deprotonation and rearrangement. The crystal structure of the rearranged trimer derivative has been published [173].

$$ (30) $$

Phosphazene   Phosphorin

7 Reactions of Main Group Organometallic Reagents with Cyclophosphazenes

The reactions of main group organometallic reagents with cyclophosphazenes are complex and lead to a large variety of products [8]. This has been ascribed to the participation of different reaction pathways such as metal-halogen exchange, proton abstraction from the pendant organic group, ring degradation reactions and the well known nucleophilic substitution reactions. Thus the exact course of the reaction seems to depend on several factors such as the nature of the organometallic reagent, phosphazene ring size, nature of the substituent present on the phosphazene, etc. In general, the reactions of organometallic reagents with chlorocyclophosphazenes are more complex than with fluorocyclophosphazenes. Due to this, early studies in this area were performed mainly on $N_3P_3F_6$.

7.1 Reactions with Organo Lithium Reagents

In general, organo lithium reagents have been widely used to synthesise organocyclophosphazenes of the formulae, $N_nP_nX_{2n-1}R$ and *gem*-$N_nP_nX_{2n-2}R_2(X = F, R = $ alkyl or aryl). However, the complete replacement of all halogens is difficult. It appears that the electron releasing organo substituents preferentially shift the ring nitrogen lone pair electron density towards the PF_2 center thus making them inert for the subsequent nucleophilic substitutions.

Alkyl lithium reagents such as MeLi [174, 175], $CH_2=CH-CH_2Li$ [176], $PhC\equiv CLi$ [177, 178], $CH_2=C(OR)Li$ (R = Me or Et) [179] react with $N_3P_3F_6$ in a geminal manner. This can be rationalised on electronic grounds; σ-electron releasing organic substituent will tend to labilize the geminal P–X bond. But the preference of nongeminal pattern with PhLi [180], $(H_3C)_3CLi$ [181], o-MeC$_6$H$_4$Li [180, 182, 183], $LiC\equiv C$-SiMe$_3$ [178] and $C_6H_3D_2Li$ [184] suggests the operation of steric effects also as most lithium reagents are present in 'associated form' in diethyl ether. With the tetramer $N_4P_4F_8$, methyl lithium reacts geminally and the third methyl group enters antipodally (Eq. 31) [174, 175, 185].

$$N_4P_4F_8 \xrightarrow{\text{MeLi}} 2,2\text{-}N_4P_4F_6Me_2 \xrightarrow{\text{MeLi}} 2,2,6\text{-}N_4P_4F_5Me_3 \qquad (31)$$

The control of the substituents already present in the cyclophosphazene ring vs the incoming nucleophile in determining the stereochemistry of the resulting product has been observed in the aminolysis reactions. Allen and coworkers have pointed out that this is also applicable to reactions involving organometallic reagents [179]. Thus dimethylamine as an incoming nucleophile prefers to attack $\equiv PF_2$ centre rather than a $\equiv P(F)(Me)$ centre. Dimethylamino group as a π-donating substituent directs the incoming organometallic reagent to the nongeminal position (Eqs. 32, 33).

$$N_3P_3F_5[C(OEt)=CH_2] \xrightarrow{\text{HNMe}_2} 2,4\text{-}N_3P_3F_4(NMe_2)[C(OEt)=CH_2] \qquad (32)$$

$$N_3P_3F_5(NMe_2) \xrightarrow{\text{LiC(OEt)=CH}_2} 2,4\text{-}N_3P_3F_4(NMe_2)[C(OEt)=CH_2] \qquad (33)$$

It has been shown by various methods that the $N_3P_3X_5$ moiety in $N_3P_3X_5R$ exhibits a strong electron withdrawing effect [186, 187]. Thus the organic group R would be highly polar and succeptible to anionic attack, and subsequent decomposition reactions (Eqs. 34, 35) [188].

$$N_3P_3F_5(Me) \xrightarrow{\text{MeLi}} N_3P_3F_5(CH_2^- Li^+) \longrightarrow \text{Decomposition products} \qquad (34)$$

$$(35)$$

Use of an electron rich alkyl lithium reagent eliminates these side reactions by counterbalancing the electron withdrawing effect of $N_3P_3X_5$ unit. This has been demonstrated with $LiC(OR)=CH_2$ (Eqs. 36, 37) [179].

$$N_3P_3F_6 + LiC(OR)=CH_2 \longrightarrow N_3P_3F_5[C(OR)=CH_2] \qquad (36)$$

$$N_3P_3F_6 + 2LiC(OR)=CH_2 \longrightarrow gem\text{-}N_3P_3F_4[C(OR)=CH_2] \qquad (37)$$

The degradation reactions are also avoided when acidic hydrogens are not present in the organic substituent [177,178]. Thus, $N_3P_3F_6$ reacts with $PhC\equiv CLi$ in diethyl ether to give $N_3P_3F_5(C\equiv CPh)$ and gem-$N_3P_3F_4(C\equiv CPh)_2$ at the mono and bis stages respectively [178]. In general, ring degradation reactions have been reported to occur predominantly in the reaction of alkyl or aryl lithium reagents with chlorocyclophosphazenes [189,190]. However, recently Van de Grampel and coworkers have reported that the reaction of $N_3P_3Cl_6$ with MeLi under controlled conditions at low temperatures leads to the hydridophosphazenes, [191–193] gem-$N_3P_3Cl_4(H)(OPr^i)$ after work up with isopropanol (Eq. 38).

$$ (38) $$

Also, Lithio carboranes react with $N_3P_3Cl_6$ to give the corresponding monosubstituted derivatives (Eq. 39) [194, 195].

$$ (39) $$

$$R = CH_3 \text{ ; } Ph$$

Reactions of lithio metallocenes with $(NPX_2)_{3,4}(X = F, Cl)$ have been studied extensively [196–203]. The reactions of lithio ferrocenes or lithio ruthenocenes with fluorocyclophosphazenes are cleaner than the corresponding reactions with chlorocyclophosphazenes. Mono lithio metallocenes react with $N_3P_3F_6$ in a nongeminal fashion owing to the steric bulk of the metallocenyl groups while with tetramer, $N_4P_4F_8$, the second metallocene enters antipodally (Schemes 13, 14) [196–200]. Dilithio metallocenes give ansa and transannular 2,6-substituted products. However, dilithio ruthenocene gives, in addition to the above products, an intermolecularly bridged product. Interestingly, a bis trans-annular product has been isolated from the reaction of $N_4P_4F_8$ with two equivalents of dilithio ruthenocene (Scheme 14) [200]. The reactions of lithio metallocenes with chlorocyclophosphazenes are hampered by the involvement of metal-halogen exchange and ring contraction reactions. Low yields and large number of products are a characteristic feature of these reactions [201–203].

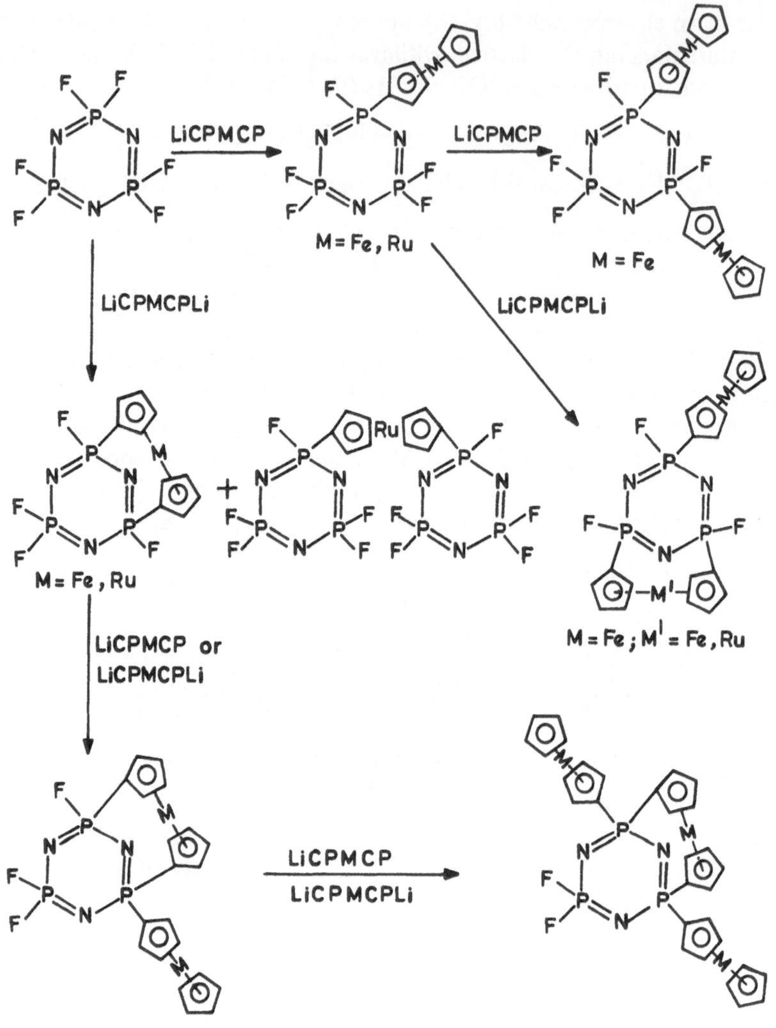

Scheme 13

The bicyclic compound $(N_3P_3Cl_4Ph)_2$ undergoes novel P–P bond cleavage on treatment with lithium triethyl borohydride to give the phosphazeno anion [204]. This species has been utilized to synthesize geminal diorganocyclo-triphosphazenes in which one organic unit is an aromatic group (Eq. 40).

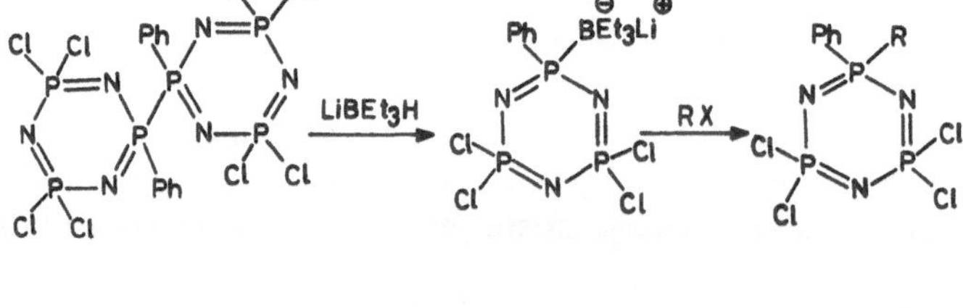

(40)

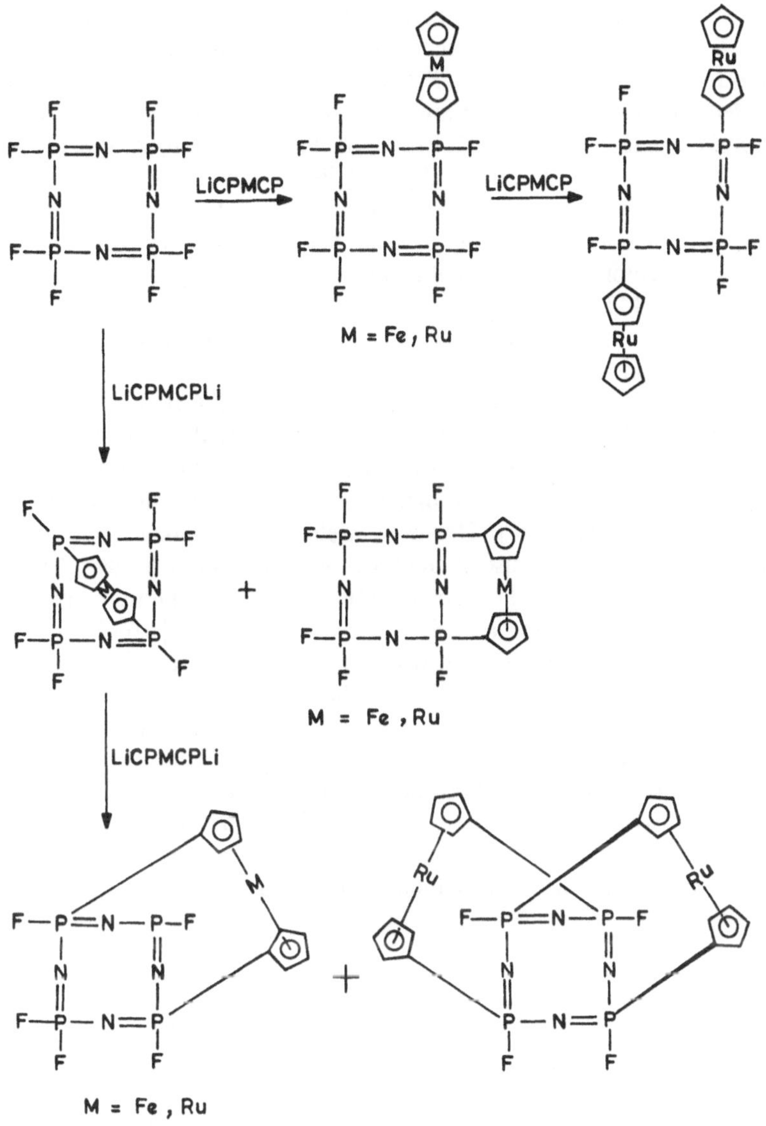

Scheme 14

7.2 Reactions with Grignard Reagents

Grignard reagents react with chlorocyclophosphazenes in a geminal pathway, to yield mono and diorgano derivatives [205–210]. However, at the monostage a P–P linked bicyclophosphazene has also been obtained [205, 206]. Bicyclophosphazene formation is restricted by the use of bulky Grignard reagents such as Pr^iMgCl, Bu^tMgCl, $(CH_3)_3SiCH_2MgCl$ [206]. But reaction of $PhMgCl$ with $N_3P_3Cl_6$ gives only the bicyclophosphazene derivative $(N_3P_3Cl_4Ph)_2$ (Scheme 15). The interaction of $N_4P_4Cl_8$ with Grignard reagents results in ring contracted products [211–213]. Shaw proposed a mechanism to account for this ring contraction (Scheme 16) [211, 212]. The initial attack to RMgX breaks the P–N skeleton to form a mono alkylated linear chain compound in which magnesium is covalently bound to an imino nitrogen. Then the 1,6-ring closure reaction with a concomitant removal of MgXCl gives the ring contracted product. Later chlorine replacement reactions occur exclusively at the exocyclic PCl_2R centre [214].

Recently Allcock and coworkers have synthesised nongeminal di and tri-organo cyclotriphosphazenes in an indirect method starting from nongem-$N_3P_3(NMe_2)_n(Cl_{6-n}$ (n = 3 or 4) [146]. Alkylation with Grignard reagents and elimination of NMe_2 group by treatment with HCl gives the nongeminal organo derivatives (Scheme 17). This procedure, however, suffers from the ether cleavage reactions [146, 215] which ultimately lead to the formation of alkoxycyclophosphazenes (Scheme 17). However, yields of organocyclophosphazenes have been optimized by the use of an excess of the Grignard reagent. Use of the copper complex, $[Cu(PBu_3)I]_4$ in conjunction with the Grignard reagent maximizes the formation of diorgano derivatives (Scheme 18) [216–222]. Van de Grampel and coworkers have recently reassigned the structure of the intermediate involved in these reactions [223]. They have also shown that these 'copper-magnesio' intermediates react with aldehydes and ketones to generate phosphazenes containing pendant hydroxy functionality (Scheme 18)

The reaction of $N_3P_3Cl_6$ with trimethyl aluminium yields the persubstituted derivative, $N_3P_3Me_6$ and a ring opened linear phosphazene salt (Eq. 41) [223, 225]. The reaction was found to proceed via both geminal and nongeminal substitution pathways, with the geminal pathway predominating. The degree of

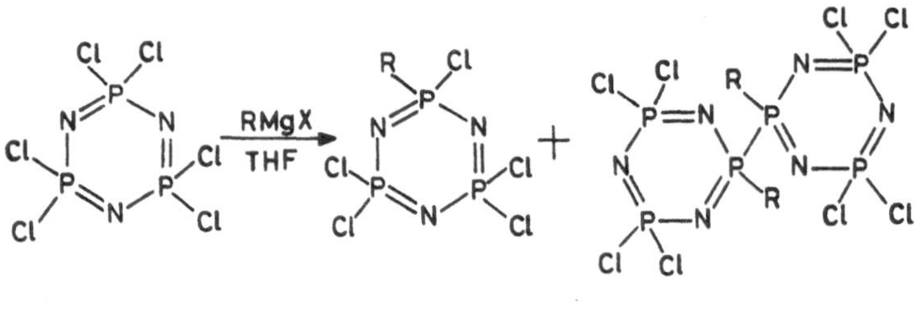

PhMgCl	0 %	100 %
MeMgCl	15 %	85 %
n-BuMgCl	69 %	31 %
$(CH_3)_3SiCH_2MgCl$	100 %	0 %

Scheme 15

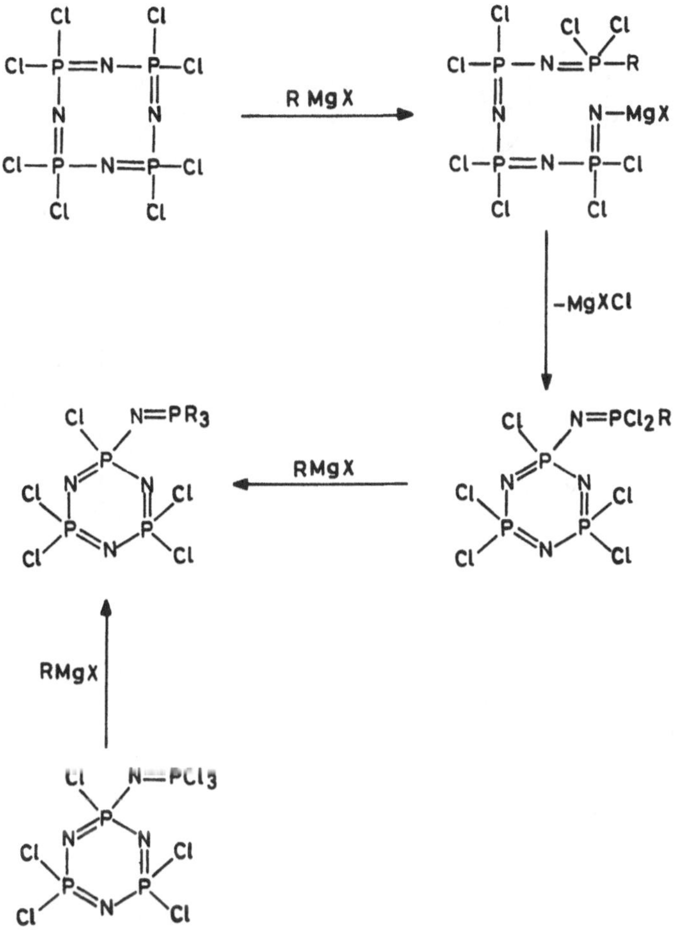

Scheme 16

ring opening decreases with increasing methyl substitution of the phosphazene ring.

$$N_3P_3Cl_6 + AlMe_3 \longrightarrow$$

$$N_3P_3Me_6 + [Me_2P=N-P(Me)_2=N-P(Me)_2=NH_2^+]Cl^- \qquad (41)$$

7.3 Reactions of Organophosphazenes with Alkyl Lithium Reagents

As pointed out earlier (Sect. 7.1), alkyl lithium reagents induce metal-hydrogen exchange reactions. This possibility was investigated first by Paddock and coworkers [266, 227]. They have found that the anion generated by the reaction of methylphosphazene with n-butyl lithium interacts with electrophiles such

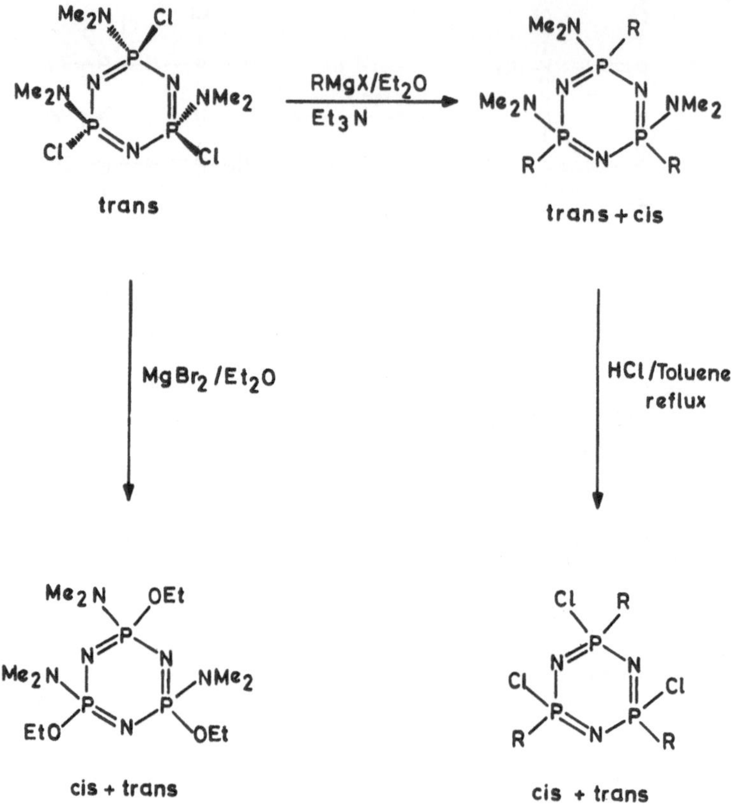

Scheme 17

as $(CH_3)_3SiCl$, $(CH_3)_3GeCl$, $(CH_3)_2SiCl_2$, C_6H_5COOEt, etc. Representative conversions are illustrated in Scheme 19. Recently Allcock and co-workers have extended these reactions to the trifluoroethoxy phosphazenes, $N_3P_3O(Ph)_5(OCH_2CF_3)$ [228]. The metal-hydrogen exchange at the α-carbon atom of the OCH_2CF_3 group results an anion which has been treated with nucleophiles such as CH_3I and Ph_3SnCl (Scheme 20).

7.4 Friedel-Crafts Reactions

In contrast to the organo lithium and Grignard reagents Friedel-Crafts arylation reactions occur quite cleanly with $N_3P_3Cl_6$, while unsubstituted fluorocyclophosphazenes do not undergo Friedel-Crafts arylation at all [1, 13]. Friedel-Crafts arylation predominantly results in geminal products irrespective of the starting cyclophosphazene (Eq. 42) [229]:

$$N_nP_nCl_{2n} \xrightarrow{\text{ArH}} gem\text{-}N_nP_n(Ar)_m(Cl_{2n-m}(n = 3, 4)(m = 2, 4, 6) \qquad (42)$$

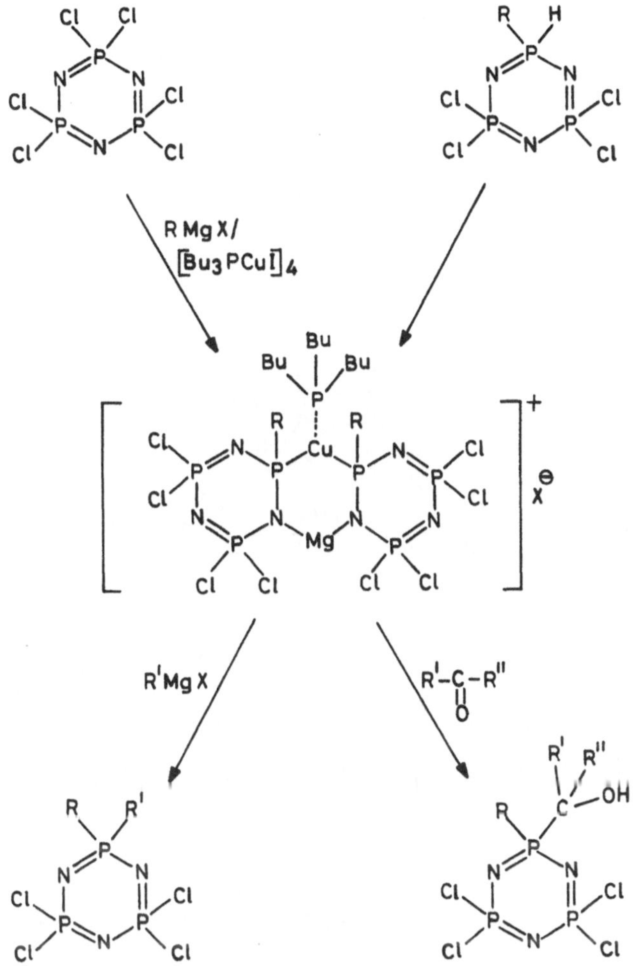

Scheme 18

An extensive study of the Friedel-Crafts phenylations of a wide variety of aminophosphazenes have been conducted [230–233]. Reactions at the PCl_2 centre is generally sluggish. With the bulky secondary amino derivatives, hydrocarbon formation is the competing process [232]. Aryl fluorophosphazenes undergo Friedel-Crafts arylation readily [182, 234–236]. These reactions have been used to convert a P(Ar)F centre to a P(Ar)(Ar') centre (Eqs. 43, 44) [234].

$$N_3P_3F_5(C_6H_3D_2) \xrightarrow[\text{AlCl}_3/\text{Et}_3\text{N}]{C_6D_6} gem\text{-}N_3P_3F_4(C_6H_3D_2)(C_6D_5) \qquad (43)$$

$$2,4\text{-}N_3P_3F_4(C_6H_3D_2)_2 \xrightarrow[\text{AlCl}_3/\text{Et}_3\text{N}]{C_6D_6} gem\text{-}N_3P_3F_2(C_6H_3D_2)_2(C_6D_5)_2 \qquad (44)$$

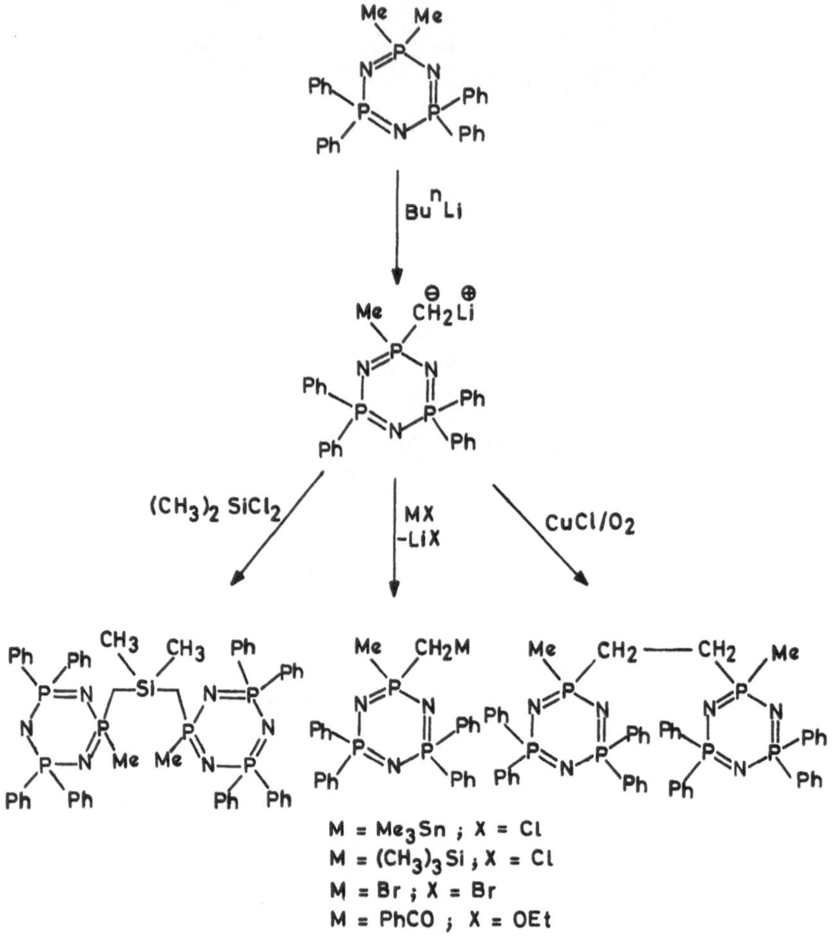
Me Me
Ph Ph Ph Ph
Bu^nLi
Me CH_2Li
$(CH_3)_2SiCl_2$
MX
-LiX
CuCl/O2
CH3 CH3
Me CH2M
Me CH2 — CH2 Me
M = Me3Sn ; X = Cl
M = (CH3)3Si ; X = Cl
M = Br ; X = Br
M = PhCO ; X = OEt

Scheme 19

8 Hydridophosphazenes: Syntheses and Reactivity

Hydridophosphazenes have been obtained according to the following reaction (Eq. 45) [237, 238]:

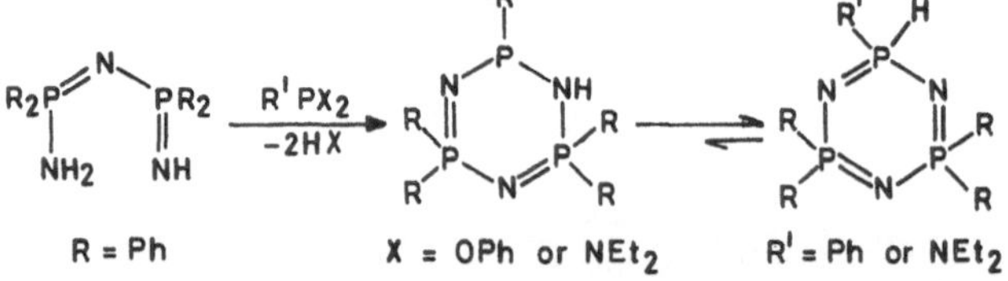
R'PX2
-2HX
R = Ph
X = OPh or NEt2
R' = Ph or NEt2

(45)

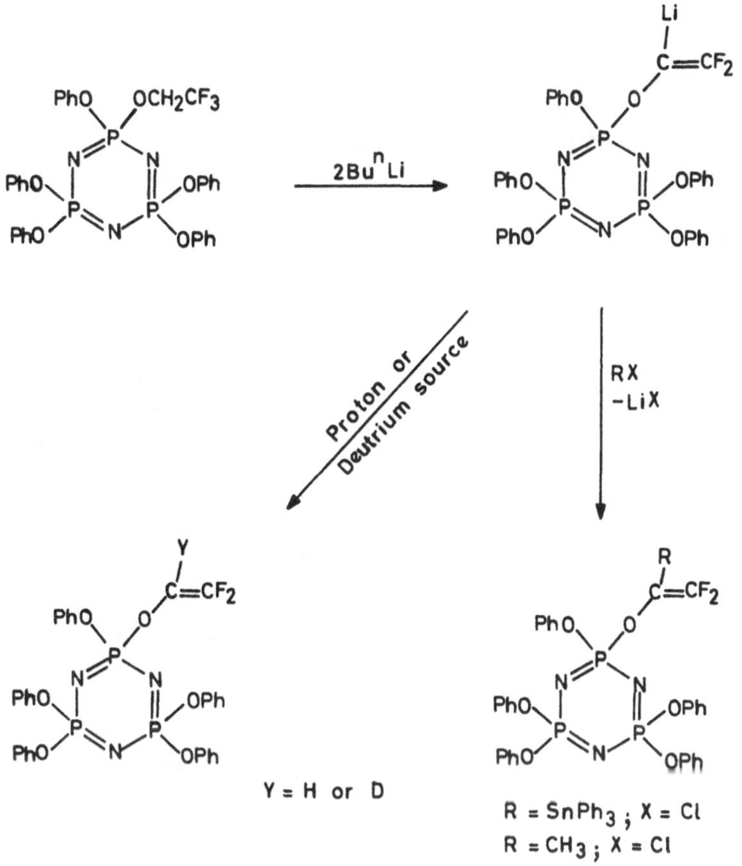

Scheme 20

The initially formed cyclotriphosphazadienes rapidly tautomerise to result in hydridocyclotriphosphazenes. Alternatively, the hydridophosphazenes are accessible from the interaction of metallophosphazene intermediates with proton sources (Eqs. 39, 46) [216, 217, 191–193].

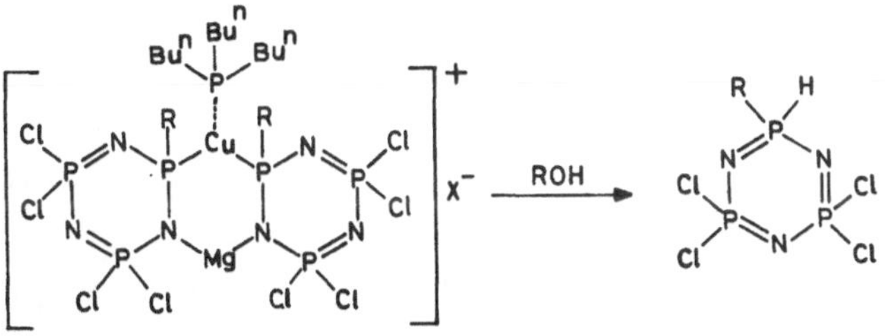

$$(46)$$

Presence of a polar phosphorus-hydrogen bond makes hydridophosphazenes as versatile reagents to generate a large variety of phosphazene derivatives. In particular the insertions of P–H bond into carbon-heteroatom double bonds have been explored in great detail (Scheme 21) [239]. Addition to thiocyanates occurs regiospecifically at the C = N centre. Similar reactions occur with olefins; in the resulting product, the phosphorus is attached to the electron deficient carbon centre (Scheme 21) [239].

Chlorophosphines, R_2PCl, $RPCl_2$ (R = Me, etc.) react with hydridophosphazenes to afford the hydrochloride salts of phosphinophosphazenes. From the latter, neutral cyclophosphazenes can be liberated on deprotonation with a base. The hydrochloride adducts of phosphinocyclotriphosphazenes undergo numerous reactions. These are illustrated in Scheme 22 [240]. One important observation is that on heating they lead to the formation of a dimeric phosphazene, linked through a phosphorus–phosphorus single bond (Scheme 22)

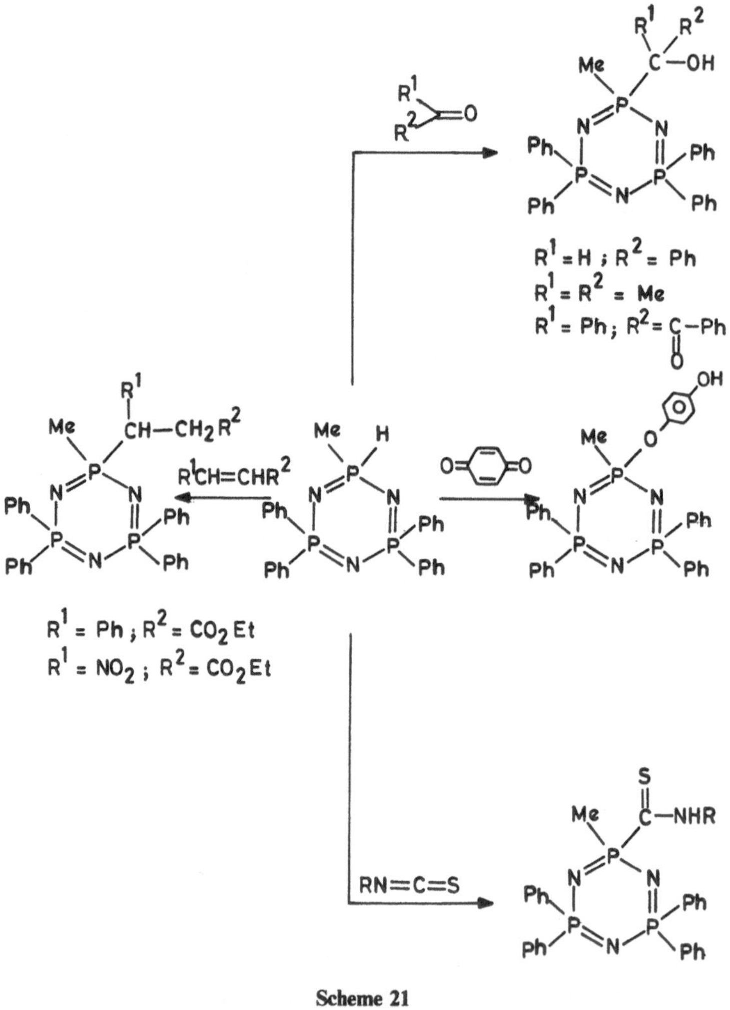

Scheme 21

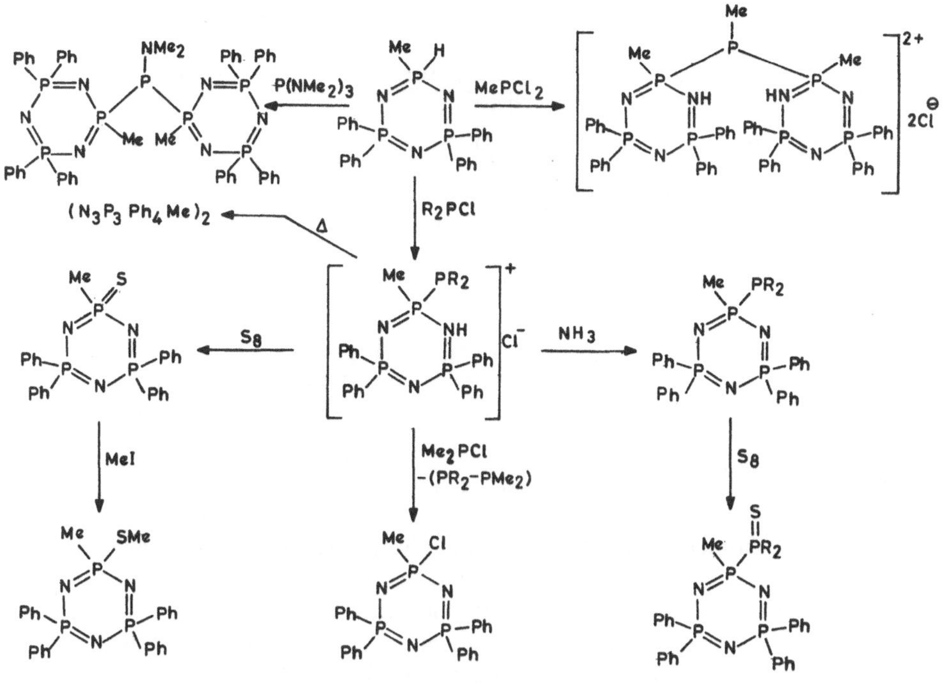

Scheme 22

[241]. Methyl phosphido linked dimeric phosphazene derivative $\{[N_3P_3Ph_4(Me)H]_2P(Me)\}^{2+}2Cl^-$ has been obtained when $MePCl_2$ is allowed to react with the hydridophosphazene, $N_3P_3Ph_4(Me)H$ (Scheme 22) [240]. Oxidations of the hydridophosphazenes with $KMnO_4$, sulphur or halogen lead to the corresponding phosphoryl [238], thiophosphoryl [242] or halophosphazene derivatives [243, 244, 218, 216] (Eq. 47) The thiophosphoryl derivative has also been obtained by the reaction of the hydrochloride salt of the corresponding phosphinophosphazene with sulphur (Scheme 22) [240].

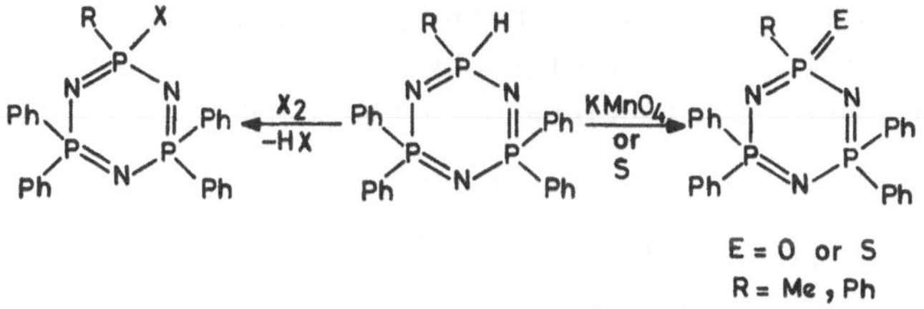

$$E = O \text{ or } S$$
$$R = Me, Ph$$

(47)

Alkyl lithium reagents react with the hydridophosphazenes to generate phosphazeno anions [241]. These anions undergo facile reactions with non-metal halides or alkyl halides to produce phosphazenes with the phosphorus

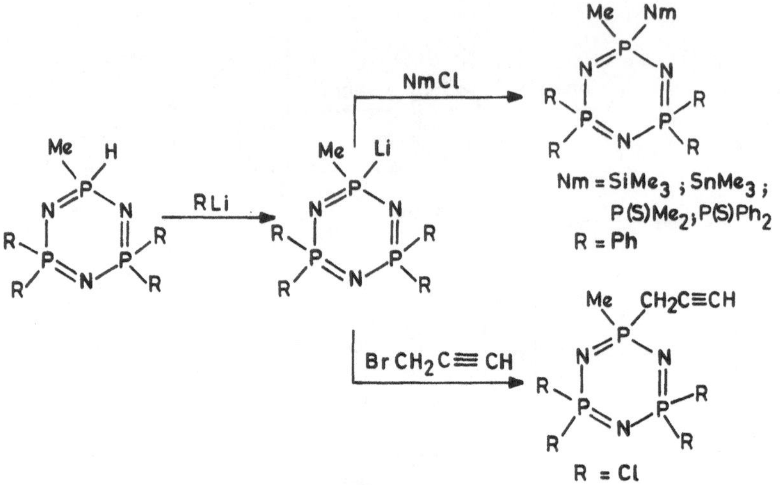

Scheme 23

attached to non-metals such as silicon, tin, phosphorus or carbon (Scheme 23) [245].

The interaction of hydridophosphazenes with Grignard reagents in the presence of monoalkyl pentachlorocyclotriphosphazene, $N_3P_3Cl_5R$ leads to novel bicyclic products $(N_3P_3Cl_4R)_2$ (Eq. 48). With bulky R substituents such as n-butyl and i-propyl groups no bicyclic products are formed [220, 221].

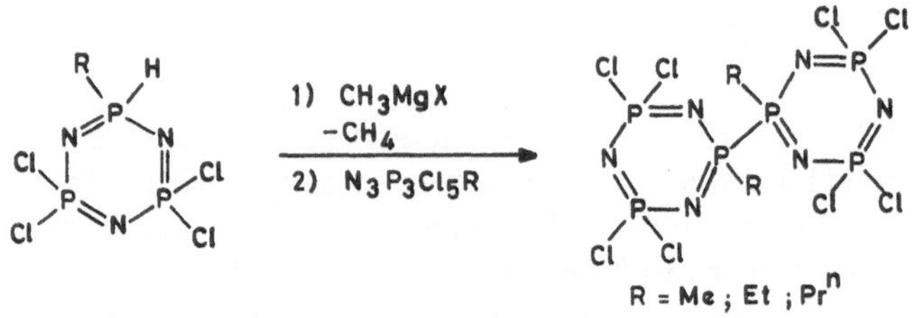

$$(48)$$

In all the above reactions the hydrido phosphorus centre acts as a pentavalent phosphorus. Interesting reactions are known in which the tautomer of the hydridophosphazene containing a P(III) centre acts as a trivalent phosphine towards transition metals. Thus $N_3P_3Ph_4(Me)H$ on treatment with MCl_2 (M = Pd, Pt) [246] or $AuCl_3$ [247] results in complexes $[N_3P_3Ph_4(Me)H]_2$ MCl_n(M = Pd or Pt; n = 2, M = Au; n = 3). These complexes have been shown to possess phosphorus-metal coordinate bonds (Eq. 49).

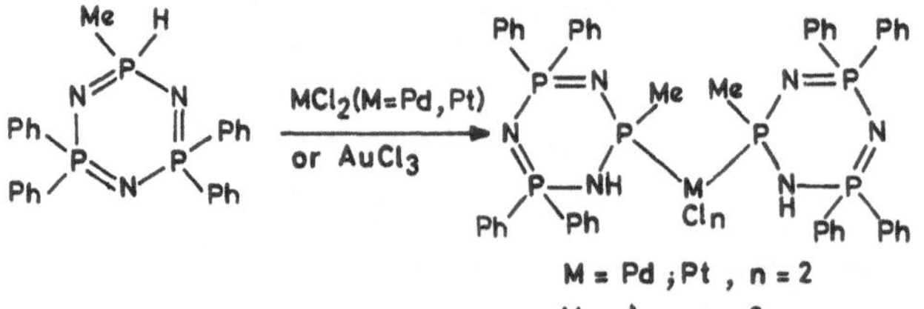

$$(49)$$

9 Interaction with Transition Metal Ions

Although most of the chemistry of cyclophosphazenes has been centred on the reactions at phosphorus, there have been consistent attempts to explore the coordination chemistry of these compounds [7, 8]. Since the ring nitrogen atoms of the cyclophosphazenes possess a lone pair of electrons these can be viewed as potential donor sites towards suitable groups including transition metal ions. Alternative strategies of using the cyclophosphazenes as ligands have focussed on the incorporation of suitable exocyclic groups possessing donor sites. This strategy allows for the synthesis of a variety of ligands and also serves as model studies towards the design of polyphosphazene ligand systems. Although suggestions have been made for the involvement of the ring π-skeleton in coordination [248], no examples analogous to the borazine complex, $B_3N_3Me_6 \cdot Cr(CO)_3$ are known.

9.1 Ring Nitrogen Coordination

Although the parent halogenocyclophosphazenes themselves are known to be poor ligands and form only ionic salts such as $N_3P_3Cl_5^+ AlCl_4^-$ or adducts such as $N_3P_3Br_6 \cdot AlCl_3$, substitution of the halogens by electron releasing groups such as alkyl or amino substituents renders the ring nitrogen atoms more basic. Additionally, while the six membered rings are rigid and therefore, are poor ligands due to their structural inflexibility, increasing the ring size enhances ring puckering and hence preferential coordination geometries become more feasible. Thus, while $N_3P_3(NMe_2)_6$ forms only ring protonated species [250, 251] $[N_3P_3(NMe_2)_6H^+]_2X^{2-}$; $X^{2-} = CoCl_4^{2-}$ or $Mo_6O_{19}^{2-}$, $N_4P_4Me_8$ forms a complex with cupric chloride where one ring nitrogen atom is protonated while the opposite ring nitrogen is coordinated to a $CuCl_3$ unit [252]. Both $N_4P_4Me_8$ and $N_4P_4(NHMe)_8$ form neutral 1:1 complexes with $PtCl_2$ [253–255]. In both of these examples platinum is present in a square planar geometry. Coordination is through two antipodal ring nitrogen atoms in a *cis* geometry and two chloride ligands. Interestingly, ionic salts of the type $[N_4P_4Me_8H^+]_2MCl_4^{2-}$ M = Co or Pt are also formed [254, 256].

Higher membered ring systems are more facile in their reactions towards transition metal ions. $N_5P_5Me_{10}$ forms a protonated complex $[N_5P_5Me_{10}H_2]^{2+}[CuCl_4]^{2-}$ and neutral complexes $N_5P_5Me_{10} \cdot M(CO)_3$ (M = Mo, W) [257, 258]. The twelve membered ring system $N_6P_6Me_{12}$ coordinate in a *cis* manner through two ring nitrogen atoms to $PdCl_2$ and $PtCl_2$ (Fig. 7) [259]. However, another twelve membered ring system $N_6P_6(NMe_2)_{12}$ coordinates through four ring nitrogen atoms to copper and cobalt [260–262]. In these examples the coordination around the metal is distorted trigonal bipyramidal (Fig. 7). Four (ring) nitrogen atom coordination is also observed in

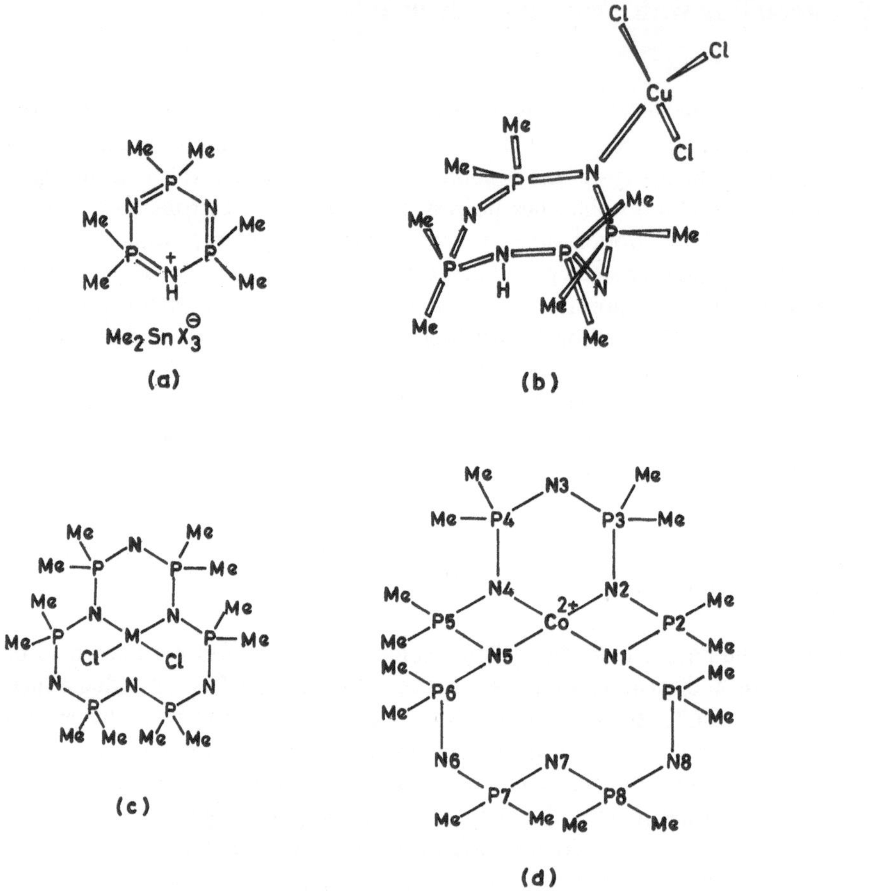

Fig. 7a–d. Structures of (a) $[N_3P_3(Me)_6H]^+[Me_2SnX_3]^-$ (b) $N_4P_4(Me)_8H \cdot CuCl_3$ (c) N_6P_6 $(Me)_{12} \cdot MCl_2$ (M = Pt or Pd) and (d) $[N_8P_8(Me)_{16}Co]^{2+}(NO_3)_2$

the complexes of the sixteen membered ring $N_8P_8Me_{16}$ [263], as for example in the complex $[N_8P_8Me_{16} \cdot Co(NO_3)]^+NO_3^-$.

9.2 Exocyclic Group Coordination and Direct Ring Phosphorus Interaction with Transition Metals

Several exocyclic groups with different coordination capabilities have been incorporated using the well studied substitution reaction pathways in halogenocyclophosphazenes. These have included (a) nitrogen containing groups such as pyrazolyl [264], Schiff bases [265] (Eqn. 50, 51) (b) acetylinic units [177] (c) phosphino units directly linked to the ring phosphorus atom [266] (Eq. 52) (d) phosphino groups attached through a spacer group [267]

(Eq. 53) (e) carboranyl substituents [194, 195] etc. (Eq. 39). There have also been reports on the incorporation of cyclophosphazenes in macrocycles such as phthalocyanins [268] and porphyrins [269–271].

$$N_3P_3R_2Cl_4 + PzH \xrightarrow{Et_3N} N_3P_3R'_2(Pz)_4 \qquad (50)$$

R = R' = Ph, R = Cl(R' = Pz = 3,5-dimethyl pyrazole)

$$N_3P_3Cl_6 \xrightarrow[\text{(ii) } H_2\backslash Pd]{\text{(i) } p\text{-}HOC_6H_4NO_2} N_3P_3(OC_6H_4N{=}CHAr)_6 \qquad (51)$$
$$\text{(iii) ArCHO} \qquad (Ar = \text{aromatic, pyridyl etc.})$$

$$[N_3P_3Cl_4Ph]_2 \xrightarrow[\text{2) PPh}_2Cl]{\text{1) LiBEt}_3H} 2N_3P_3Cl_4Ph(PPh_2) \qquad (52)$$

$$N_3P_3(OPh)_5Cl \xrightarrow[\text{(ii) Bu}^n Li\backslash PPh_2Cl]{\text{(i) } p\text{-}BrC_6H_5OH} N_3P_3(OPh)_5(O{-}C_6H_4PPh_2) \qquad (53)$$

Acetylinic phosphazenes $N_3P_3F_5(C{\equiv}CPh)$ [177], $2,2\text{-}N_3P_3Cl_4R(CH\,C{\equiv}CH)$ and $2,2\text{-}N_3P_3Cl_4R(C{\equiv}CMe)$ form π-complexes with dicobalt octa carbonyl with the loss of two carbon monoxide molecules [272]. The acetylinic unit functions as a four electron donor. No participation from the ring nitrogen atom is observed. Similarly the phosphino derivatives $N_3P_3(OPh)_5(OC_6H_4PPh_2)$ and $N_3P_3(OC_6H_4PPh_2)_6$ form complexes with a variety of substrates such as AuCl, $H_2Os_3(CO)_{10}$, $Mn(CO)_3(\eta^5\text{-}C_5H_5)$, $Fe(CO)_3[PhCH{=}CHC(O)Me]$, $[RhCl(CO)_2]_2$ etc [267]. Only pendant phosphines participate in coordination. Cyclophosphazenes containing a phosphine attached directly to the ring phosphorus atom $2,2\text{-}N_3P_3Cl_4Ph(PPh_2)$ forms complexes with $Cr(CO)_6$, $Fe_2(CO)_9$ or $Ru_3(CO)_{12}$ [266]. In the former two examples mononuclear complexes are isolated in which the phosphino phosphorus replaces a CO from the metal. In the latter case the trinuclear core of the metal frame work is retained. Similarly carboranyl phosphazenes synthesized from the reaction of *nido*-carborane anion with chlorocyclophosphazenes also form complexes with $Rh(PPh_3)_3Cl$, $Mo(CO)_6$ or $W(CO)_6$ [273].

Aminophosphazenes contain both exocyclic nitrogen atoms as well as ring nitrogen atoms. Basicity studies on a series of these type of compounds have clearly revealed that the endocyclic nitrogens are more basic[3]. This has been verified in the coordination behaviour of several aminocyclophosphazene derivatives (vide supra). Very few examples are known where exocyclic nitrogens also participate in coordination. In one such example, $N_4P_4(NMe_2)_8$ forms a complex with $W(CO)_6$ (Eq. 54) [274].

$$N_4P_4(NMe_2)_8 + W(CO)_6 \longrightarrow N_4P_4(NMe_2)_8{\cdot}W(CO)_4 \qquad (54)$$

The coordination around tungsten is completed by an exocyclic dimethyl-amino nitrogen and a ring nitrogen. More recently several pyrazolyl cyclophosphazenes, $N_3P_3(Pz)_6$, $N_3P_3Ph_2(Pz)_4$, $N_3P_3Ph_4(Pz)_2$ [264, 275],

$N_3P_3(NMeCH_2CH_2O)_2 (Pz)_4$ [276] have been synthesized. The last example is a spirocyclic derivative formed from N-methyl ethanolamine. In these cases the coordination response of the ligands varies from metal to metal. Thus $N_3P_3(Pz)_6$ coordinates exclusively via geminal pyrazolyl pyridinic nitrogens to $PtCl_2$ or $PdCl_2$ [264, 275] while two exocyclic pyrazolyl pyridynic nitrogens and one ring nitrogen atoms are involved in coordination to $CuCl_2$; the geometry around copper is found to be distorted trigonal bipyramidal [277]. Again, $N_3P_3Ph_4(Pz)_2$ and $N_3P_3(NMeCH_2CH_2O)_4(Pz)_2$ react with $Mo(CO)_6$ by the loss of three carbon monoxide molecules [276]. Coordination to molybdenum is now through two geminal exocyclic pyrazolyl nitrogens and one ring nitrogen atom (Scheme 24). Clearly, more studies are needed to rationalise these divergent coordination responses.

An interesting ligating behaviour of aryl groups attached to cyclophosphazenes has been recently reported (Eqs. 55, 56) [278].

$$N_3P_3F_5R + Cr(CO)_6 \longrightarrow N_3P_3F_5R \cdot Cr(CO_3) \tag{55}$$

$$R = C_6H_5, OC_6H_5 \text{ etc.}$$

$$N_3P_3Cl_6 + Na[RCr(CO)_3] \longrightarrow N_3P_3Cl_{6-n}[RCr(CO)_3]_n \tag{56}$$

$$R = OC_6H_5 \text{ etc., n} = 1 \text{ or } 6.$$

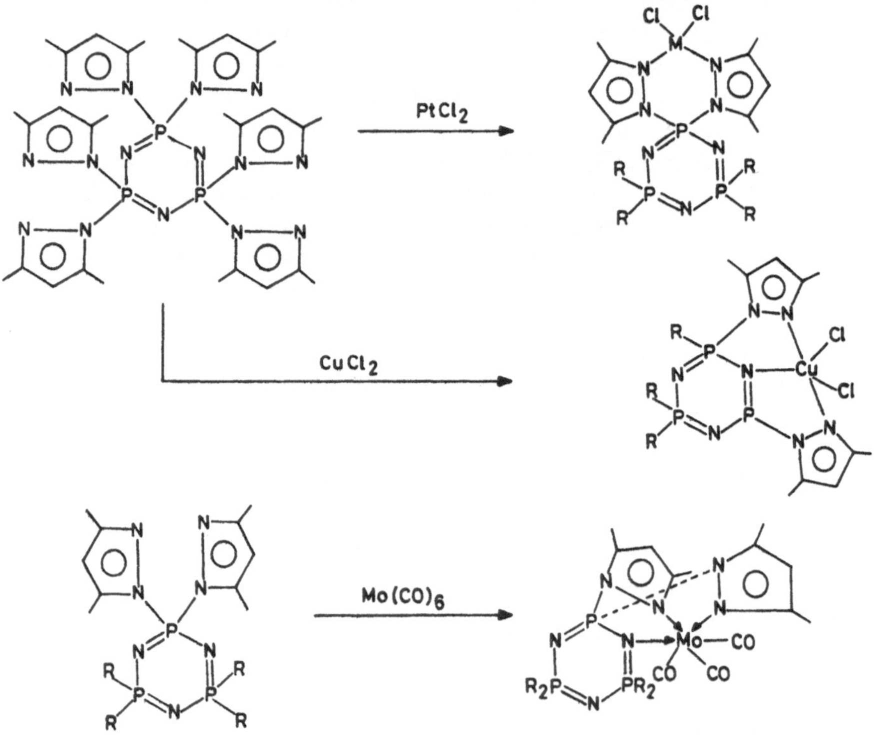

Scheme 24

In the first method the phenyl or phenoxy substituent on $N_3P_3F_5R$ serves as a π-type $6e^-$ donor and coordinates to $Cr(CO)_3$. In the latter method a preformed phenol or its derivative first forms a η^6-complex with $Cr(CO)_3$. This reagent subsequently is reacted with the halogenocyclophosphazene affording high metal loadings.

Several examples of direct ring phosphorus interaction with transition metals are now known [279–287]. For example, reaction of $N_3P_3Cl_6$ with $Na_2Fe_2(CO)_8$ affords a novel spirocyclic diiron octa carbonyl derivative (Scheme 25) [282, 283]. The diiron spiro derivative acts as a template for the construction of several transition metal clusters. Some of the other examples where ring phosphorus atom is involved in interaction with transition metals are summarized in Scheme 25. The ring phosphorus atoms in hydrido phosphazenes, $N_3P_3R_4R'H$ also coordinates to transition metals. This has been discussed in an earlier section (vide supra).

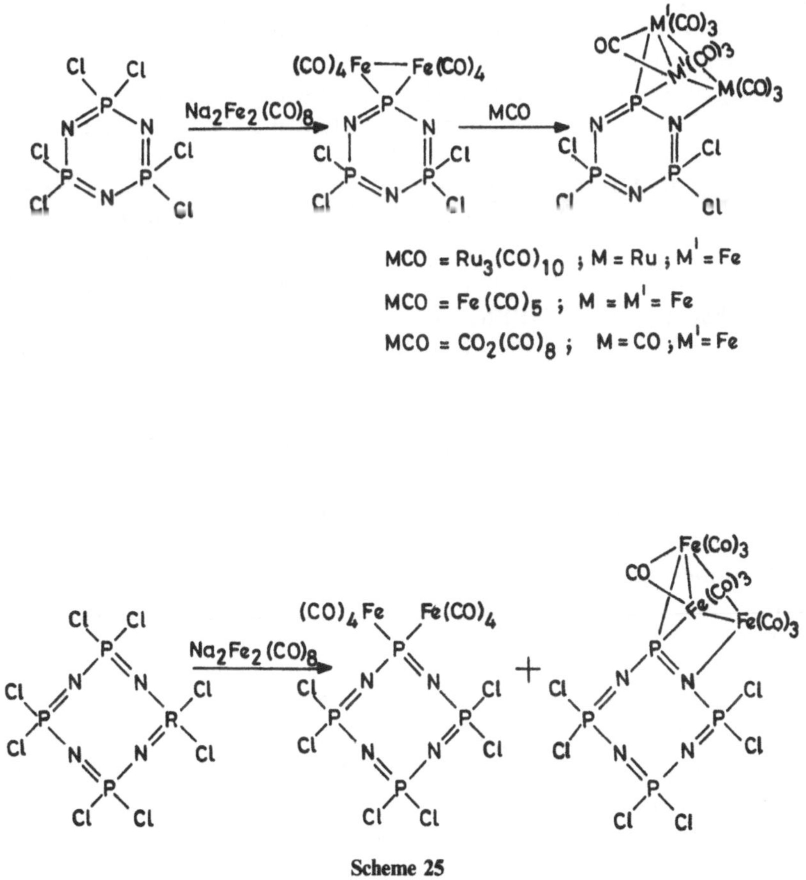

$MCO = Ru_3(CO)_{10}$; $M = Ru$; $M' = Fe$

$MCO = Fe(CO)_5$; $M = M' = Fe$

$MCO = CO_2(CO)_8$; $M = CO$; $M' = Fe$

Scheme 25

10 Physical Methods

10.1 Proton NMR

Proton NMR has been extremely valuable in the structural elucidation of cyclophosphazene derivatives containing alkylamino, alkoxy or simple alkyl substituents [287].

The three bond coupling constant between phosphorus and proton often observed in cyclophosphazenes is designated as $^3J^*$(P–H) and is actually a composite value of ^3J(P–H) and $^5J^*$(P–H). However, the latter value is often very small and can be neglected [288]. Generally coupling constants involving geminal substitution [ex. $\equiv P(OCH_3)_2$] are smaller (9–14 Hz) than coupling constants involving nongeminal substitution [ex. $\equiv P(Cl)(OCH_3)$] (13–20 Hz) [3]. This effect though not universal has been observed in many instances. Thus, in the assignment of structures of products formed in the reactions of phosphazenyl pentachlorocyclophosphazene, $N_3P_3Cl_5NPPh_3$ with amines or alcohols, the coupling constant differences have been found to be very useful (Scheme 26) [289–294]. For the derivative $N_3P_3Cl_4(NPPh_3)$ (NMe_2) [291], three isomers are possible, A(geminal), B(nongeminal '*cis*') and C(nongeminal '*trans*'). While the N–CH$_3$ resonance is a doublet at 2.63 ppm for A with a coupling constant [$^3J^*$(P–H)] of 13.5 Hz in the anologous nongeminal derivatives B and C the observed coupling constant increases to 16.3 Hz. Similar enhancement of the value of $^3J^*$(P–H) is observed in the other examples. Interestingly the NPPh$_3$ group has a strong shielding effect on its *cis* neighbours. It can be seen that in the nongeminal derivatives $N_3P_3Cl_4(NPPh_3)$ (NMe_2) (B and C), $N_3P_3Cl_3(NPPh_3)$ $(NMe_2)_2$(D and E), $N_3P_3Cl_4(NPPh_3)$ (OCH_3) (F and G) and $N_3P_3Cl_4(OCH_3)_2$(H and I) the resonances of N–CH$_3$ or O–CH$_3$ that are *cis* to NPPh$_3$ are more shielded than the corresponding resonances in the *trans* isomer. Derivatisation has been successfully used for structure elucidation where proton NMR spectra of the parent chloro alkylamino or alkoxycyclophosphazenes are ambiguous [3]. Thus the structure of the bis (*N*-methyl anilino) derivative, 2,6-$N_4P_4Cl_6[N(Me)C_6H_5]_2$ has been identified by studying the proton NMR of its methoxy derivative 2,6-$N_4P_4(OMe)_6[N(Me)C_6H_5]_2$ [63]. If the structure of the parent chlorocyclophosphazene derivative is 2,6-*trans*, the corresponding methoxy derivative should exhibit two OCH$_3$ resonances in a 1:2 intensity. However, for a 2,6-*cis* structure three methoxy resonances in a 1:1:1 intensity would be seen. The observed spectrum agrees with the expectation for the 2,6-*trans* isomer. However, this method should be used cautiously in light of recent geminal-nongeminal transformations studied by Shaw and coworkers in the alkoxylation of 2,2-$N_3P_3Cl_4(NH_2)_2$ (vide supra) [168–170].

Often the proton NMR spectra in cyclophosphazenes is complicated by the presence of second order effects in the form "virtual coupling" [3, 82]. This is manifested in the form of a broad hump that often is seen between the normal fine structure due to the first order spectrum. The criterion for the observation of

δ N–CH$_3$: 2.63(13.5) δ N–CH$_3$: 2.52(16.3) δ N–CH$_3$: 2.64(16.5)

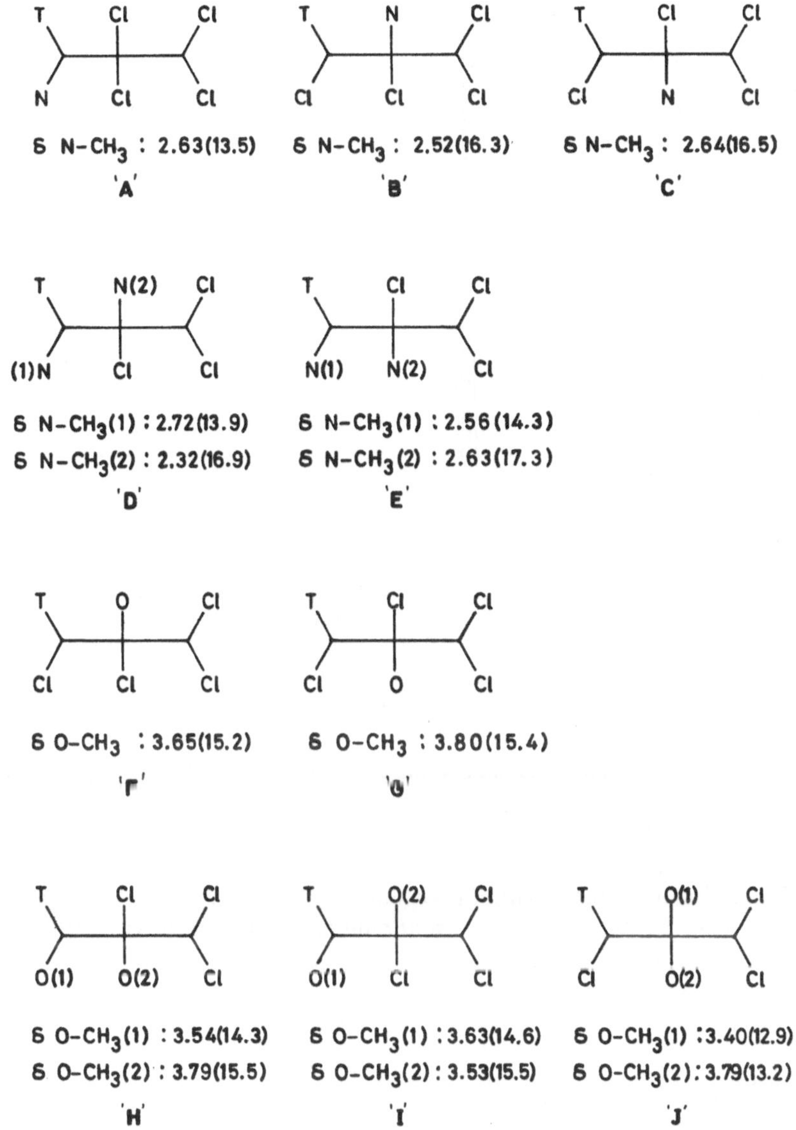

'A' 'B' 'C'

δ N–CH$_3$(1) : 2.72(13.9) δ N–CH$_3$(1) : 2.56(14.3)
δ N–CH$_3$(2) : 2.32(16.9) δ N–CH$_3$(2) : 2.63(17.3)

'D' 'E'

δ O–CH$_3$: 3.65(15.2) δ O–CH$_3$: 3.80(15.4)

'F' 'G'

δ O–CH$_3$(1) : 3.54(14.3) δ O–CH$_3$(1) : 3.63(14.6) δ O–CH$_3$(1) : 3.40(12.9)
δ O–CH$_3$(2) : 3.79(15.5) δ O–CH$_3$(2) : 3.53(15.5) δ O–CH$_3$(2) : 3.79(13.2)

'H' 'I' 'J'

Scheme 26

the virtual coupling are as follows: (1) $^3J^*(P–H) \geqslant {}^2J(P–P)$ (2) the chemical shift differences between the phosphorus nuclei involved should be very small.

Manytimes the presence or absence of virtual coupling can be used for structural assignment. Thus for the spirocyclic derivative $N_3P_3Cl_4$ [NH(CH$_2$)$_2$NH] [96], the chemical shift differences between PCl$_2$ and P$_{spiro}$ are extremely small; in fact only a single resonance is observed, 23.3 ppm when the spectrum was recorded at 27.5 MHz. Due to this the $^3J^*(P–H)$ is larger than

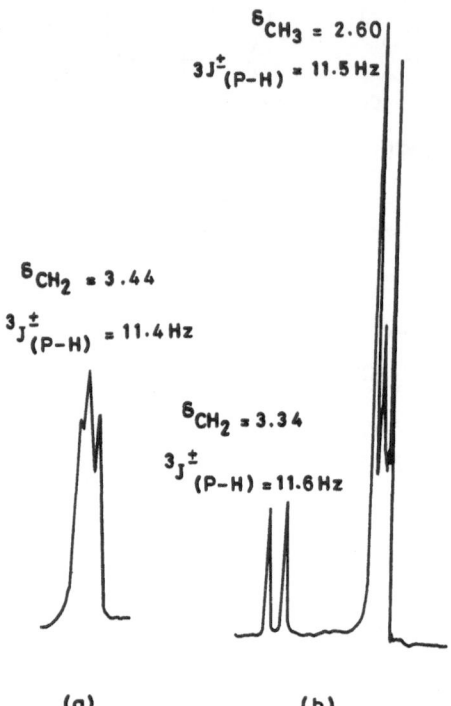

$\delta_{CH_3} = 2.60$

$3J^{\pm}_{(P-H)} = 11.5\,Hz$

$\delta_{CH_2} = 3.44$

$3J^{\pm}_{(P-H)} = 11.4\,Hz$

$\delta_{CH_2} = 3.34$

$3J^{\pm}_{(P-H)} = 11.6\,Hz$

Fig. 8. ^1H NMR spectra of $N_3P_3Cl_4$ [NHCH$_2$CH$_2$NH] (a) and $N_3P_3(NMe_2)_4$ [NHCH$_2$CH$_2$NH] (b) (from Ref. 82)

(a)   (b)

either $^2J(P-P)$ or the chemical shift differences between PCl$_2$ and P$_{spiro}$. Therefore a virtual coupling is seen in the proton NMR. Replacement of chlorines by N(Me)$_2$ affords $N_3P_3(NMe_2)_4$[NH(CH$_2$)$_2$NH] [295]. Here now the chemical shift differences between P(NMe)$_2$ and P$_{spiro}$ are larger than the $^3J^*(P-H)$ for N–CH$_2$; a clean doublet without virtual coupling is seen for the N–CH$_2$ resonance (Fig. 8). But the chemical shift difference between the two P(NMe$_2$)$_2$ nuclei are very small (close to zero) in comparison with the $^3J^*(P-H)$ for N–CH$_3$. This results in a virtual coupling for the N–CH$_3$ resonance (Fig. 8).

10.2 ^{31}P NMR Spectroscopy

^{31}P NMR spectroscopy has been widely used to assign the structures of the cyclophosphazene derivatives. ^{31}P NMR data for selected cyclophosphazene derivatives are given in Tables 3–5. The chemical shift of a particular phosphorus atom depends on several factors such as, nature of the substituent (bulkiness and electronegativity), extent of π-bonding with the exocyclic substituent, exocyclic and endocyclic bond angles, etc. Substituents elsewhere on the ring may also influence the chemical shift of a phosphorus nuclei under consideration. Thus, the chemical shift of the PCl$_2$ group in $N_3P_3Cl_6$ [296] is 19.3 ppm while, in gem-$N_3P_3(NPPh_3)_2Cl_4$ it is shifted upfield to 13.4 ppm

Table 3. ^{31}P NMR Spectral data for selected persubstituted cyclotri and tetraphosphazenes[a]

compound	chemical shift (ppm)	compound	chemical shift (ppm)
$N_3P_3Cl_6$	+ 19.3	$N_4P_4Cl_8$	− 6.7
$N_3P_3F_6$	+ 13.9	$N_4P_4F_8$	− 17.0
$N_3P_3(NHEt)_6$	+ 18.0	$N_4P_4(NHEt)_8$	+ 4.3
$N_3P_3(NHMe)_6$	+ 21.5	$N_4P_4(NHMe)_8$	+ 11.1
$N_3P_3(NHBu^i)_6$	+ 18.1	$N_4P_4(NHBu^i)_8$	− 3.1
$N_3P_3(NMe_2)_6$	+ 24.6	$N_4P_4(NMe_2)_8$	+ 9.6
$N_3P_3(OPh)_6$	+ 9.9	$N_4P_4(OPh)_8$	+ 12.6
$N_3P_3(OC_6H_4Me\text{-}o)_6$	+ 8.8	$N_4P_4(OC_6H_4Me\text{-}o)_8$	− 14.5
$N_3P_3(OC_6H_4Ph\text{-}o)_6$	+ 6.7	$N_4P_4(OC_6H_4Ph\text{-}o)_8$	− 14.2
$N_3P_3Ph_6$	+ 15.2	$N_4P_4Ph_8$	+ 5.2
$N_3P_3(NC_2H_4)_6$[b]	+ 37.0	$N_3P_3(Im)_6$[b]	− 2.2
$N_3P_3(OMe)_6$	+ 21.6	$N_3P_3(SMe)_6$	+ 45.7
$N_3P_3(OCH_2CH_3)_6$	+ 18.3	$N_3P_3(OCH_2=CH_2)_6$	+ 11.3
$N_3P_3[NMe(CH_2)_2NMe]_3$	+ 29.4	$N_3P_3[NH(CH_2)_3NH]_3$	+ 18.6
$N_3P_3[O(CH_2)_2O]_3$	+ 37.4	$N_3P_3[O(CH_2)_3O]_3$	+ 14.1
$N_3P_3[O(CH_2)_4O]_3$	+ 21.7		

[a] Chemical shift values are relative to 85% H_3PO_4 and upfield shifts are negative.
[b] NC_2H_4 = Aziridinyl, Im = Imidazolyl

[297]. Similarly alkoxy and aryloxy substituents in the neighbouring phosphorus exert a downfield shift. It has been observed that certain substituents such as Br, Ph, etc. shield the phosphorus by a neighbouring group anisotropy effect.

On increasing the ring size from trimer, $N_3P_3R_6$ to tetramer, $N_4P_4R_8$ a large upfield shift is noticed (Table 3). The extent of this upfield shift is further influenced by the nature of substituents. Thus when R = Ph, 10 ppm upfield shift is observed while the electronegative fluorine causes a 30.9 ppm upfield shift.

^{31}P NMR can be used to identify positional isomers. The two positional isomers resulting from $N_3P_3(NPPh_3)(OMe)_4X$ are differentiated by using the ^{31}P NMR spectra [289, 290]. Thus, the derivative 2,46,6,2,4-$N_3P_3(OMe)_4$ $(NPPh_3)Cl$ where chlorine and $NPPh_3$ occupy nongeminal cis-positions has three different ring phosphorus nuclei and an ABX pattern is observed. In the gem-$N_3P_3(OMe)_4(NPPh_3)F$ where $NPPh_3$ and fluorine occupy geminal positions only two distinct ring phosphorus nuclei are present (Fig. 9). Similarly the structures of the disubstituted tetrameric isomers 2,4-$N_4P_4X_6R_2$ and 2,6-$N_4P_4X_6R_2$ can easily be assigned with the aid of ^{31}P NMR [29]. The 2,6-derivative, in principle, should give a A_2B_2 or A_2X_2 pattern ^{31}P NMR spectra. But in most cases a symmetrical triplet is observed owing to the accidental proximity of the chemical shifts of the nuclei. However, the compound 2,6-$N_4P_4Cl_6(NC_2H_4)_2$ gives a clear A_2X_2 type ^{31}P NMR spectrum [29]. The 2,4-derivative would result in a AA'BB' or AA'XX' spectrum. However, isochronous phosphorus-31 chemical shifts can be misleading and a careful analysis is often required before unambigous structural assignments are made.

Table 4. ^{31}P NMR Spectral data for the compounds of general formula $N_3P_3RR'Cl_4$[a]

No	$N_3P_3RR'Cl_4$		δPCl_2 (ppm)	$\delta PRR'$ (ppm)	$^2J_{PP}$ (Hz)	Ref.
	R	R'				
1	NH_2	Cl	22.2	18.9	49.0	170
2	NHEt	Cl	20.3	18.7	43 ± 2	37
3	$NHPr^i$	Cl	20.8	17.5	46.0	37
4	$NHCH_2Ph$	Cl	21.3	18.1	46.6	52
5	NHPh	Cl	21.2	11.7	47.1	39
6	$NHBu^t$	Cl	16.0	5.3	40.6	296
7	NMe_2	Cl	20.5	21.6	49.1	43
8[b]	$N(CH_2Ph)_2$	Cl	19.6			52
9	NC_2H_4	Cl	22.2	31.2	39.0	29
10	NC_5H_{10}	Cl	20.8	18.7	48.0	44
11	$NPPh_3$	Cl	20.3	0.2	47.5	293
12	OMe	Cl	22.5	16.7	63.3	289
13	OCH_2CF_3	Cl	16.3	22.8	66.0	73a
14	$OCH=CH_2$	Cl	23.4	13.2	64.0	299
15	$O(CH_2)_3OH$	Cl	23.5	15.9	62.1	139
16	OPh	Cl	23.2	13.2	60.9	301
17	$OC_6H_4Me\text{-}p$	Cl	22.5	12.2	60.0	79
18	$OC_6H_4Cl_2\text{-}o$	Cl	21.0	12.5	60.0	73b
19	SEt	Cl	21 ± 2	41 ± 2	ng	81a
20	NH_2	NH_2	21.1	8.2	50.0	170
21	$NHCH_2Ph$	$NHCH_2Ph$	21.0	9.4	44.4	52
22	$NHSiMe_3$	$NHSiMe_3$	22.1	9.4	50.9	298
23	NHPh	NHPh	20.4	2.3	48.0	39
24	$NHBu^t$	$NHBu^t$	19.6	2.3	44.7	298
25	$NPPh_3$	$NPPh_3$	13.4	-10.9	35.8	297
26	NC_2H_4	NC_2H_4	21.9	34.2	30.0	29
27	$NPPh_3$	NH_2	16.9	-1.6	43.5	297
28	$NPPh_3$	NHMe	18.2	0.8	41.2	291
29	$NPPh_3$	$NHBu^t$	15.9	-5.9	42.7	291
30	$NPPh_3$	NMe_2	16.6	3.9	36.8	291
31	$NPPh_3$	NEt_2	16.0	1.0	37.4	291
32	$NPPh_3$	NC_2H_4	18.1	10.3	34.2	291
33	$NPPh_3$	NC_5H_{10}	16.6	1.9	35.3	291
34	OPh	OPh	24.4	0.6	64.1	301
35	$OCH=CH_2$	$OCH=CH_2$	24.5	-0.6	68.8	300
36	$OC_6H_4Cl_2\text{-}o$	$OC_6H_4Cl_2\text{-}o$	22.8	-1.3	70.0	73b
37	SMe	SMe	23 ± 2	50 ± 2	ng	80
38	SEt	SEt	17.7	51.7	4.8	81a
39	H	Me	17.6	13.8	12.0	216, 217
40	H	Et	18.4	20.4	8.0	216, 217
41	H	Bu^t	18.4	32.3	u	216, 217
42	H	OPr^i	20.9	2.5	26.0	191
43	D	Me	18.6	11.4	12.0	216, 217
44	Me	Me	18.0	35.7	8.9	206–208
45	Me	Et	18.3	41.8	< 2	206–208
46	Me	CH_2COMe	19.2	31.4	9.7	206–208
47	Me	Bu^t	18.1	50.4	< 2	206–208
48	Me	$OCH=CH_2$	21.6	29.6	21.9	299, 300
49	Et	Et	19.2	48.1	< 2	206–208
50	Ph	Ph	17.1	19.5	12.0	296

[a] legend: ng = not given in the original reference, u = unresolved
[b] single line observed due to accidental isochronous chemical shift

Table 5. ^{31}P NMR spectral data of selected spiro cyclotriphosphazene derivatives[a]

No	compound	n = 2			n = 3			n = 4			Ref
		δPCl_2	δP_{spiro}	$^2J_{PP}$	δPCl_2	δP_{spiro}	$^2J_{PP}$	δPCl_2	δP_{spiro}	$^2J_{PP}$	
1	$N_3P_3[NH(CH_2)_nNH]Cl_4$	23.5	22.9	47.1	21.5	7.5	45.5	20.82	13.07	46.0	89, 90
2	$N_3P_3Ph_2[NH(CH_2)_nNH]Cl_4$[b]	22.8	25.4	30.1	20.4	10.9	22.0	21.4	14.5	32.0	93
3	$N_3P_3[NH(CH_2)_nNH]Cl_2$	–	–	–	23.1	12.3	43.7	–	–	–	140
4	$N_3P_3[NH(CH_2)_nNMe]Cl_4$	23.8	20.4	41.8	21.7	16.5	6.5	–	–	–	92
5	$N_3P_3Ph_2[NMe(CH_2)_nNMe]Cl_2$[c]	24.0	22.9	25.2	–	19.4	53.9	–	–	–	93
6	$N_3P_3[NMe(CH_2)_nNMe]_2Cl_2$	29.1	24.9	54.3	–	–	–	–	–	–	140
7	$N_3P_3[O(CH_2)_nNH]Cl_4$	24.9	24.3	53.9	22.4	7.2	53.8	–	14.3	–	140
8	$N_3P_3[O(CH_2)_nNMe]Cl_4$	25.1	22.4	53.9	–	–	–	–	–	–	90
9	$N_3P_3[O(CH_2)_nNH]_2Cl_2$	29.0	29.0	52.74	23.8	12.5	33.0	–	–	–	140
10	$N_3P_3Ph_2[O(CH_2)_nNH]Cl_2$[d]	23.7	26.4	39.3	20.6	10.0	–	22.9	16.9	44.2	93
11	$N_3P_3[O(CH_2)_nNMe]_2Cl_2$	30.7	28.5	62.9	–	–	–	–	–	–	90
12	$N_3P_3Ph_2[O(CH_2)_nNMe]Cl_2$[e]	24.4	25.2	35.7	–	–	–	–	–	–	93
13	$N_3P_3[O(CH_2)_nO]Cl_4$	26.5	24.5	68.0	24.1	3.4	69.2	24.1	10.3	70.5	139
14	$N_3P_3Ph_2[O(CH_2)_nO]Cl_2$[f]	25.2	26.7	50.3	21.9	5.4	45.1	23.6	12.5	53.9	93
15	$N_3P_3[O(CH_2)_nO]_2Cl_2$	31.3	30.95	76.8	26.5	9.1	70.8	27.8	16.0	76.9	139

[a] chemical shift values are quoted in ppm and coupling constants have a unit Hz
[b] δPPh_2 values are: 20.3 (n = 2), 20.0 (n = 3), 18.7 (n = 4).
[c] δPPh_2 values are: 20.8 (n = 2), 20.1 (n = 3).
[d] δPPh_2 values are: 20.8 (n = 2), 21.1 (n = 3), 20.7 (n = 4).
[e] δPPh_2 value is: 22.1 (n = 2).
[f] δPPh_2 values are: 23.1 (n = 2), 22.1 (n = 3), 22.0 (n = 4).

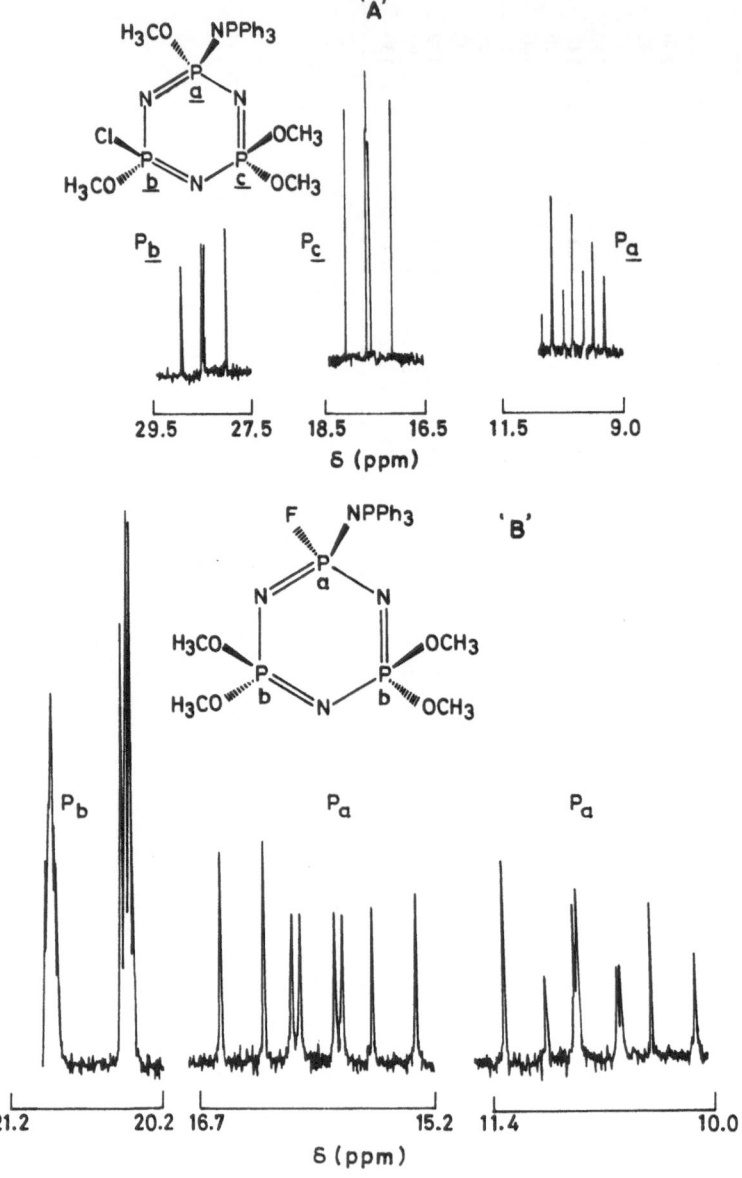

Fig. 9a, b. $^{31}P\{^1H\}$ NMR spectra of (a) $N_3P_3NPPh_3(OMe)_4Cl$ and (b) $N_3P_3NPPh_3(OMe)_4F$ (from Ref. 289)

Two bond phosphorus-phosphorus coupling constants are also found to be informative in the identification of geminal and nongeminal isomers in the case of tetrameric derivatives. In geminal derivatives, $N_4P_4R_2Cl_6$ or $N_4P_4R_4Cl_4$ the coupling constant $^2J(PR_2–N–PCl_2)$ is small when compared to the coupling constant $^2J(PRCl–N–PCl_2)$ in the corresponding nongeminal derivatives.

For example, the $^2J(PR_2$–N–$PCl_2)$ value for the geminal isomer of $N_4P_4Cl_6(NC_2H_4)_2$ is 11.6 Hz, while the analogous coupling constant $[^2J(PRCl$–N–$PCl_2)]$ in the nongeminal 2-*cis*-4 isomer is 27.1 Hz, thus allowing for an easy distinction between the two isomers [29]. In the geminal and nongeminal 2-*cis*-4 isomers of formula $N_4P_4(NC_2H_4)_2Cl_6$ are 11.6 Hz and 27.1 Hz respectively.

Empirical relationships have been proposed to correlate $^2J(P$–$P)$ with electronegativity of the substituents on the phosphorus atoms concerned [302, 303]. However, several other factors including the stereochemical dispositions of the substituents seems to affect the magnitude of the coupling constants. Recently, four bond phosphorus–phosphorus coupling constants have been measured for a number of phosphazenyl phosphazenes and their magnitude appears to have significance in relation to the conformation of the phosphazenyl group ($N=PPh_3$) relative to the phosphazene ring [304].

10.3 Other Spectroscopic Techniques

The Fluorine-19 NMR spectroscopy has been used to confirm the structures of fluoroalkylamino, fluoroalkyl and fluoroalkoxycyclophosphazenes. The fluorine chemical shifts (relative to $CFCl_3$) span a wide range from -18.0 ppm to -71.9 ppm^3. It has been observed that the magnitude of $^4J(P$–$F)$ is sensitive to the *cis* and *trans* orientation of the relevant P–F bonds. Thus, in general, in *cis* isomers $^4J(P$–$F)$ is *ca.* -1.0 Hz while, in the *trans* isomer it is larger ($\sim +12.0$ Hz) [25].

The most important feature in the infrared spectra of cyclophosphazenes is a very strong broad band in the region 1150–1450 cm^{-1} attributable to a degenerate P–N ring stretching vibration [25]. The value of the $v_{P=N}$ increases with increasing electronegativity of the substituents on phosphorus. In contrast, the electron releasing substituents such as alkylamino tend to weaken the skeletal π-bonding thereby reducing the ring P=N stretching frequency [305]. Thus, for $N_3P_3F_6$ $v_{P=N}$ appears at 1300 cm^{-1} while, for $N_3P_3(NMe_2)_6$ it is seen around 1195 cm^{-1}. In addition to a high frequency shift of a ca. 40–60 cm^{-1}, splitting of the $v_{P=N}$ band is also observed when the ring nitrogen is protonated or involved in coordination to a metal [305–308]. This is a consequence of the delocalisation of the nitrogen lone pair electron density to the metal rather than to the phosphazene ring.

Even small differences in the P–Cl bond lengths of chlorocyclophosphazenes give rise to a multiple line ^{35}Cl NQR spectrum [309–314]. An increase in the P–Cl bond length is accompanied by a decrease in the ^{35}Cl NQR frequency, and this most probably reflects an increased ionic character of the P–Cl bonds. Closely related chlorocyclophosphazenes show a linear relationship between ^{35}Cl NQR frequencies and P–Cl bond lengths [314].

10.4 X-ray Diffraction Studies

A brief discussion of the generally accepted model of bonding present in cyclophosphazenes is relevant for rationalising the X-ray structural data. For detailed theories in this regard the reader is referred to earlier reviews and other articles [1-3, 9]. The basic σ-frame work of the ring system is constituted from the sp^3 hybrid orbitals of phosphorus and approximately sp^2 hybrid orbitals of nitrogen. Two sp^2 orbitals are involved in bonding to phosphorus while the third which is in-plane with the P–N–P segment is occupied by a lone pair. The odd electron on nitrogen resides in the p orbital (p_z, if the ring skeleton is assumed to be in the XY plane) perpendicular to the plane of the ring, while the one on phosphorus is present on any appropriate orbital such as $3d$, $4s$, $4p$ or their combinations. Three types of π-bonding situations have been envisaged. (a) A combination of the odd electrons on nitrogen and phosphorus to give a π_a bonding (this is antisymmetric to the P–N–P plane). (b) A combination of the inplane sp^2 orbital of nitrogen and an appropriate d orbital such as $d_{x^2-y^2}$ or d_{xy}, to give a π_s bonding (this is symmetric to the P–N–P plane). (c) a π-bonding between phosphorus and the exocyclic substituent involving a conjugative electron release from the substituent lone pair into a phosphorus d orbital. The major contribution to the ring π-bonding seems to be of the π_a type. A distinction has been made by Craig and Paddock in a π_a type [1]: if d_{xz} orbitals are involved exclusively from phosphorus it leads to a delocalized bonding with alternate signs around the ring (heteromorphic) while if d_{yz} are used exclusively it leads to constant signs of interaction (homomorphic) [1]. Dewar [1] has suggested equal contribution from d_{xz} and d_{yz} orbitals resulting in the π_a system to be localized over a P–N–P segment. It has been shown that both these approaches converge to the same result. Some amount of transannular phosphorus–phosphorus bonding has also been suggested. Several X-ray structures of cyclophosphazenes have been completed and the results correlated with the above bonding models. Structural parameters for representative examples are given in Tables 6–8. Allen has compiled a comprehensive list of X-ray structures in a recent review [1].

In many crystal structure determinations importance is given to the following aspects:

(a) variations in the P–N bond distances and ring phosphorus and nitrogen bond angles.
(b) P–N ring conformations.

In homogenously persubstituted cyclophosphazenes $N_3P_3R_6$ (R = halogen, alkylamino, alkoxy, etc.) bond distances are virtually equal within a molecule. The endocyclic phosphorus-nitrogen bond lengths fall in the range of 1.55 Å–1.61 Å. This value is much shorter than the normal P–N single bond distance and is as a result of the multiple bond character between phosphorus and nitrogen. In addition, increase of electronegativity of exocyclic substituent(R) decreases the ring P–N bond length (Table 6). The endocyclic phosphorus

Table 6. Structural parameters of persubstituted cyclophosphazenes[a]

No	Compound	Ring Conformation	Bond distances		Bond angles			Ref.
			$P=N_{endo}$	$P-X_{exo}$	$N\hat{P}N$	$P\hat{N}P$	$X\hat{P}X$	
1	$N_3P_3Cl_6$	planar	1.581	1.98	118.4	121.4	101.4(1)	315
2	$N_3P_3F_6$	planar	1.57(1)	1.52(1)	120(1)	119(1)	100(1)	316
3	$N_3P_3Br_6$	slightly chair	1.576(8)	2.162(4)	118.5(5)	121.4(5)	102.1(1)	317
4	$N_3P_3(NCS)_6$	nonplanar	1.58(5)	1.63(1)	119(3)	121(4)	100(1)	318
5	$N_3P_3(NMe_2)_6$	distorted boat	1.588(—)	1.652(4)	116.7(4)	123.0(4)	101.5(8)	319
6	$N_3P_3(NC_2H_4)_6 \cdot CCl_4$	planar	1.590(—)	1.676(3)	116.7(2)	123.3(2)	99.0(2)	328
7	$N_3P_3(Im)_6$[b]	planar	1.557	1.68	119.1	120.1	102.3	321
8	$N_3P_3[NH(CH_2)_3NH]_3$	Planar	1.600(—)	1.656(4)	115.2(4)	122.4(3)	101.9(4)	84
9	$N_3P_3(OPh)_6$	nonplanar	1.575(—)	1.582(2)	117.3(3)	121.9(3)	98.0(3)	322
10	$N_3P_3(OC_6H_4Cl-p)$	planar	1.575(—)	1.579(5)	117.4(1)	121.8(4)	98.2(3)	323
11	$N_3P_3(OC_6H_3Cl_2-o)_6$	planar	1.58(2)	1.59(1)	117.7(1)	121.8(4)	98.2(3)	73b
12	$N_3P_3(OC_6H_3Me_2-o)_6$	nonplanar	1.561(—)	1.576(3)	115.8(2)	123.7(4)	99.3(2)	73b
13	$N_3P_3(O_2C_6H_4)_3$	planar	1.59(2)	1.62(1)	117(1)	122(1)	97.4(8)	324
14	$N_3P_3(1,8\text{-dioxonaphthalene})_3$	puckered boat	1.57	1.59	117.9	122.1	102.2	325
15	$N_3P_3(2,3\text{-dioxonaphthalene})_3$	planar	1.56(2)	1.61(2)	116(1)	124(1)	96(1)	326
16	$N_3P_3(2,2'\text{-dioxobiphenyl})_3$	slightly boat	1.572(0)	1.584(4)	118.5(3)	121.0(2)	102.7(3)	327
17	$N_3P_3(Me)_6$	distorted chair	1.595(—)	1.788(2)	116.7(1)	122.5(1)	102.6(2)	328
18	$N_3P_3(Ph)_6$	slightly chair	1.597	1.804	117.8	122.1	103.8	329
19	$N_4P_4Cl_8(K\text{-form})$	boat or tub	1.570(9)	1.991(4)	121.2(5)	131.3(6)	102.8(2)	330
20	$N_4P_4(NMe_2)_8$	saddle	1.58(1)	1.69(1)	120.1(5)	130.0(6)	103.8(5)	331
21	$N_4P_4(NC_4H_8)_8$[b]	saddle	1.575(0)	1.677(7)	118.8(4)	131.7(4)	101.7(3)	332, 333
22	$N_4P_4(OMe)_8$	saddle	1.57	1.58	121.0	132.0	105.5	334
23	$N_4P_4(OPh)_8$	boat	1.560	1.582	121.1	133.9	102.3	79
24	$N_4P_4(OC_6H_4Me-o)_8$	slightly puckered	1.555	1.579	120.8	138.5	103.1	79
25	$N_4P_4(OC_6H_4Ph-o)_8$	boat	1.57	1.584	120.9	129.1	105.9	79
26	$N_4P_4(Me)_8$	distorted boat	1.591(—)	1.802(6)	119.8(3)	132.0(2)	104.1(2)	335
27	$N_4P_4(Ph)_8$	distorted boat	1.590(—)	1.809(4)	119.8(1)	127.8(2)	105.1(2)	336

[a] bond lengths are given in Å units while bond angles are in degrees.
[b] Im = 1-imidazolyl, NC_4H_8 = 1-pyrrolidinyl

Table 7. Structural parameters for the compounds of general formula $N_3P_3XX'Y_4$[a]

No	Ring substituents			Ring conformation	$P=N$ bond distances a, b, c.	Bond angles			Ref.
	X	X'	Y			α, β	γ, δ	θ_1, θ_2	
1	Cl	Cl	Cl	planar	1.581 / 1.581 / 1.581	118.4 / 118.4	121.4 / 121.4	101.3 / 101.3	315
2	NH_2	F	F	puckered	1.584(3) / 1.554(3) / 1.557(3)	116.9 / 119.5	121.5 / 120.8	106.9 / 98.3	337
3	$NHPr^i$	Cl	Cl	planar	1.590(6) / 1.554(6) / 1.566(6)	117.5(3) / 119.3(3)	121.3(4) / 121.0(4)	107.6(3) / 100.3(2)	338
4	$NPPh_3$	Cl	Cl	twist boat	1.604(2) / 1.556(2) / 1.578(2)	114.4(1) / 119.8(1)	122.1(1) / 119.4(2)	107.2(1) / 100.5(1)	304
5	$NPCl_2(CpFeCp)$	Cl	Cl	ng	1.607(4) / 1.548(5) / 1.584(5)	114.1(2) / 120.6(2)	122.2(3) / 117.7(3)	ng	202
6	C—C—PhCl / $B_{10}H_{10}$	Cl	Cl	planar	1.576(3) / 1.576(3) / 1.570(3)	117.5(2) / 118.3(2)	119.0(2) / 121.3(2)	102.4(1) / 102.0(1)	194 / 195
7	Ph	Cl	Cl	nonplanar	1.589(4) / 1.557(4) / 1.579(4)	117.3(2) / 119.0(2)	121.6(2) / 120.9(2)	103.5(2) / 101.4(5)	204
8	$Ph \cdot Cr(CO)_3$	F	F	ng	1.569(5) / 1.562(5) / 1.548(5)	117.8(2) / 119.0(2)	121.3(3) / 121.5(3)	102.2(2) / 98.1(2)	278
9	CpFeCp	F	F	nonplanar	1.586(3) / 1.551(3) / 1.556(3)	116.7(1) / 119.6(1)	121.4(2) / 119.6(2)	103.0(1) / 97.8(1)	196

No.	Substituent	Substituent	X	Ring	P–N distances	angles	angles	angles	Ref.
10	OC₆H₃Cl₂-o	Cl		planar	1.579(3) 1.567(3) 1.575(3)	118.0(2) 119.4(2)	121.4(2) 120.4(2)	101.1(1) 101.4(1)	73b
11	CrCp(CO)₃	Cl		nonplanar	1.61 1.57 1.57	112.3 119.6	120.6 118.9	ng	285
12	Pr^i	H	Cl	planar	1.610(6) 1.540(6) 1.578(6)	115.6(3) 120.1(3)	122.1(4) 119.4(4)	109.0(3) 100.0(1)	339
13	Me	Me	F	nonplanar	1.601(7) 1.552(7) 1.560(6)	114.0(4) 120.0(4)	123.2(5) 119.6(5)	107.4(4) 97.2(3)	340
14	Ph	Ph	F	slightly boat	1.618(5) 1.539(5) 1.618(5)	115.5(3) 120.6(3)	120.7(2) 120.0(1)	107.9(3) 96.9(2)	341
15	Ph	Ph	Cl	slightly chair	1.615(2) 1.555(0) 1.578(1)	115.2(0) 119.7(0)	122.0(3) 119.2(1)	104.4(1) 100.3(1)	342
16	NPPh₃	Ph	Cl	nonplanar	1.625(6) 1.556(2) 1.578(2)	114.4(1) 119.6(1)	122.2(1) 119.4(2)	107.2(1) 100.5(1)	343
17	NH₂	NH₂	F	planar	1.597(5) 1.524(5) 1.564(5)	110.9 120.5	125.7 117.2	102.4 97.1	344
18	NH₂	NH₂	Cl	nonplanar	1.616(4) 1.562(2) 1.573(2)	112.4(2) 119.9(1)	122.7(1) 118.6(2)	104.4(2) 101.2(2)	304
19	NH₂	NPPh₃	Cl	planar	1.628(3) 1.547(3) 1.573(3)	110.0(1) 120.2(2)	124.4(2) 118.3(2)	105.1(2) 99.2(1)	304
20	NEt₂	NPPh₃	Cl	nonplanar	1.634 1.542 1.574	109.9 120.6	125.2 117.5	105.5 99.0	345

Table 7. Structural parameters for the compounds of general formula $N_3P_3XX'Y_4$ [a]

No	Ring substituents			Ring conformation	P=N bond distances a, b, c	Bond angles			Ref.
	X	X'	Y			α, β	γ, δ	θ_1, θ_2	
21	$NPPh_3$	$NPPh_3$	Cl	planar	1.642(5) 1.548(5) 1.575(5)	109.2(4) 120.8(3)	119.0(4) 117.5(0)	110.9(4) 99.6(1)	346
22	$NHBu^t$	$NHBu^t$	Cl	boat	1.61(1) 1.56(1) 1.58(1)	111.5(4) 119.2(5)	124.1(5) 119.1(6)	104.0(3) 99.3(2)	347
23	$NHSiMe_3$	$NHSiMe_3$	Cl	planar	1.617(5) 1.551(5) 1.591(4)	111.6(3) 120.4(3)	125.0(3) 117.4(4)	105.9(4) 99.3(1)	298
24	$OC_6H_3Cl_2\text{-}o$	$OC_6H_3Cl_2\text{-}o$	Cl	planar	1.579(3) 1.567(3) 1.578(3)	118.0(2) 119.4(2)	121.4(2) 120.4(2)	101.1(1) 101.4(6)	73b
25	$OC_6H_3Me_2\text{-}o$	$OC_6H_3Me_2\text{-}o$	Cl	nonplanar	1.588(2) 1.565(3) 1.571(3)	116.4(1) 119.1(1)	122.2(2) 120.5(2)	101.4(1) 100.8(6)	73b
26	$N_4P_4Cl_7$	CpFeCp	Cl	ng	1.594(5) 1.566(5) 1.559(6)	117.8(3) 119.0(3)	119.7(3) 121.8(4)	107.1(2) ng	202
27	PPh_2	Ph	Cl	puckered	1.615(5) 1.550(3) 1.556(4)	114.5(2) 119.6(2)	122.6(3) 120.3(2)	106.7(1) 100.8(1)	266
28	$PPh_2 \cdot Cr(CO)_5$	Ph	Cl	ng	1.605(5) 1.556(6) 1.584(7)	114.8(3) 119.2(4)	123.0(3) 119.9(5)	107.0(2) 100.1(2)	266
29	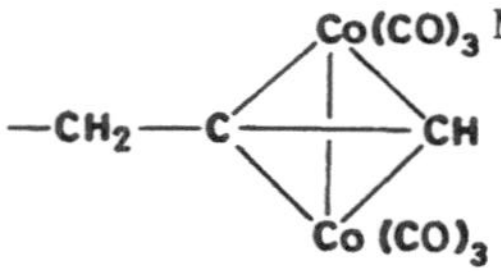 (see structure below)	Me	Cl	planar	1.606(3) 1.559(3) 1.574(3)	114.1(1) 119.9(1)	123.2(2) 119.3(2)	105.9(2) 100.2(1)	272

Structure for entry 29:

$Co(CO)_3$ — CH — $Co(CO)_3$; $-CH_2-C$

No.	X	Y	Conformation	Bond distances			Reference
30	Fe(CO)$_2$Cp	Me	ng	1.638(3) 1.546(3) 1.574(3)	111.8(1) 120.1(1)	123.7(2) 119.3(2)	110.2(1) 99.8(1)	286
31	MoCp(CO)$_3$	Cp	nonplanar	1.645 1.550 1.585	111.0 120.3	124.3 117.8	ng	285
32	FeCp(CO)$_2$	F	planar	1.666(2) 1.530(2) 1.569(3)	109.2(2) 121.4(2)	124.8(1) 117.9(2)	119.1(1) 95.4(1)	280
33	CpFeCpCH(OH)	Pr^i	envelope	1.620(2) 1.558(2) 1.585(2)	115.7(1) 119.6(1)	122.1(1) 119.7(1)	108.8(1) ng	348
34	Cl	NC$_2$H$_4$	planar	1.558(3) 1.610(3) 1.584(3)	120.9(2) 116.3(1)	121.1(2) 124.2(2)	99.7(1) ng	349
35	Cl	Ph	slightly boat	1.556(9) 1.609(9) 1.578(9)	120.7(4) 115.5(4)	121.0(5) 124.9(5)	104.4(5) 98.5(2)	350
36	NH$_2$	OMe	ng	1.612(2) 1.581(1) 1.582(1)	114.0(2) 117.0(1)	122.7(1) 122.7(1)	104.1(2) 97.1(1)	170
37	Me	Ph	nonplanar	1.599(5) 1.599(5) 1.600(5)	116.2(2) 118.5(2)	121.1(3) 120.0(2)	105.2(3) 103.3(3)	351

ᵃ bond distances are given in Å and bond angles in degrees. Legend: ng = not mentioned in the original reference,

Table 8. Bonding parameters of spiro cyclotriphosphazenes[a,b]

No	Ring substituents		Ring conformation	P=N bond distances	Bond angles			Ref.
	X–X'	Y		a, b, c	α, β	γ, δ	θ_1, θ_2	
1	NH(CH$_2$)$_3$NH	Cl	planar	1.609(5) 1.553(6) 1.579(2)	111.5(4) 119.9(2)	125.1(3) 118.3(1)	101.1(4) 99.7(2)	85, 100
2	NH(CH$_2$)$_4$NH	Cl	planar	1.595 1.573 1.565	113.5 118.2	122.8 119.0	105.1 99.6	84
3	O(CH$_2$)$_2$O	Cl	planar	1.581(2) 1.562(2) 1.571(2)	115.5(2) 119.1(1)	122.9(1) 119.6(2)	98.3(2) 100.6(1)	85
4	O(CH$_2$)$_3$O	Cl	planar	1.582(4) 1.561(4) 1.571(5)	116.6(3) 119.2(2)	122.4(3) 120.4(4)	105.4(3) 100.2(1)	85, 96
5	O(CH$_2$)$_4$O	Cl	planar	1.592(2) 1.561(2) 1.575(2)	116.1(1) 119.3(1)	122.3(1) 120.1(2)	106.1(1) 100.3(1)	96
6	Cl	MeN(CH$_2$)$_2$O	planar	1.565(4) 1.602(4) 1.579(4)	121.5(2) 114.4(2)	120.7(2) 127.0(4)	96.0(2) 98.5(1)	85, 90
7	NH(CH$_2$)$_2$NH	[NMe$_2$]$_2$	planar	1.595 1.595 1.600	115.0 116.2	124.0 122.8	95.6 100.8	352

[a] Bond distances are given in Å units and bond angles are in degrees.
[b] Crystal structures of the following compounds have also been reported: N$_3$P$_3$[O(CH$_2$)$_n$NR]Cl$_4$, R = H, n = 2, 3; R = Me, n = 2 [85], N$_3$P$_3$[NMe(CH$_2$)$_3$NMe]Cl$_4$ [85], N$_3$P$_3$(NHBu^i)$_2$[O(CH$_2$)$_3$O]Cl$_2$ [353], N$_3$P$_3$Cl$_2$[O(CH$_2$)$_2$O]$_2$ [85].

angles tend to decrease with increasing steric bulk of the substituents with the concomitant widening of the exocyclic phosphorus angle. Thus, for $N_3P_3F_6$ exocyclic and endocyclic bond angles at phosphorus are $100(1)°$ and $120(1)°$ respectively [316], while in $N_3P_3Me_6$ they are $103.8°$ and $117.8°$ respectively [328]. The presence of π-bonding between phosphorus and exocyclic nitrogen has also been inferred from X-ray crystallographic data. In the case of aminocyclophosphazenes exocyclic P–N bond lengths are in the range of 1.63 Å–1.70 Å, thus, providing evidence for the involvement of π-interaction between phosphorus and exocyclic nitrogen [1–3]. Electron withdrawing or π-conjugating substituents reduce the exocyclic P–N bond lengths. Thus, in gem-$N_3P_3Cl_4(NEt_2)$ $(NPPh_3)$, P–NEt_2 bond length is 1.64 Å (phosphazane-like nature) while the P–$NPPh_3$ bond length is diminished to 1.59 Å (close to phosphazene nature) [345]. This is rationalized in terms of increased π-bonding in the P–$NPPh_3$ segment and consequently a decreased π-bonding in the P–NEt_2 region. In the bicyclic compound, $N_4P_4(NHMe)_6(NMe)$ the bridgehead nitrogen adopts trigonal geometry and the bridge P–N bonds are phosphazane like and non-equivalent(1.729 Å and 1.709 Å) [354].

The ring conformation of cyclic trimers remain planar or slightly puckered, while the tetramers adopt several conformations owing to the enhanced flexibility of the eight membered ring. $N_4P_4Cl_8$ [330] itself crystalizes in two forms with different ring conformations; a boat (K-form) and a chair (T-form). Some tetrameric ring shapes are illustrated in Fig. 10. The energy differences among the various conformations are small. The preference of a particular conformation depends on a number of intra and intermolecular factors, such as orientation of substituents, steric and electronic nature of the substituents, crystal packing effects and hydrogen bonding interactions, etc. One fascinating example is provided by $N_4P_4Cl_4Ph_4$, where depending upon the substituent configurations the ring conformation varies. Thus in the all cis nongem-2,4,6,8-$N_4P_4Cl_4Ph_4$ the ring adopts a flattened crown conformation [355], while the 2, trans4, trans6, cis8 isomer shows a chair conformation for the ring [356]. Cyclotriphosphazenes show large deviations from planarity in the ansa and ansa-like transannular metallocenyl linked phosphazenes. In the spiro-ansa derivative, $N_3P_3Cl_2[O(CH_2)_3O]_2$ the endocyclic angle at nitrogen (which lies between the phosphorus atoms holding the ansa arch) is very small [112.7(2)°]. Again this nitrogen also lies 0.523 Å below the plane described by other five atoms of the cyclotriphosphazene ring [98].

In a heterogenously substituted compound the ring P–N bond lengths are unequal. For mono and geminal di and tetra substituted phosphazenes, $N_3P_3XX'Y_4$ there exists three pairs of P–N bonds, a, b, c. In mono or geminal disubstituted derivatives, where X = organosubstituent and X' = halogen or X = X' = organosubstituent the P–N bond lengths are in the following order: a > c > b. In geminal tetrasubstituted compounds where Y = organosubstituent the P–N bond lengths are in the order: b > c > a. Similar variations have been observed with mono spiro and dispiro derivatives also (Table 8).

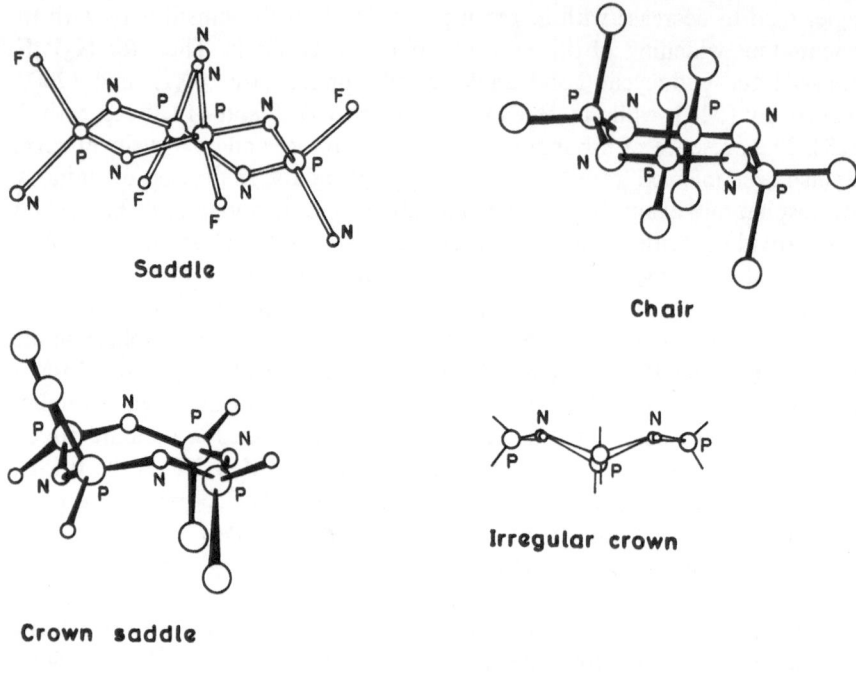

Fig. 10. Some of the ring conformations adopted by the eight membered cyclophosphazene derivatives

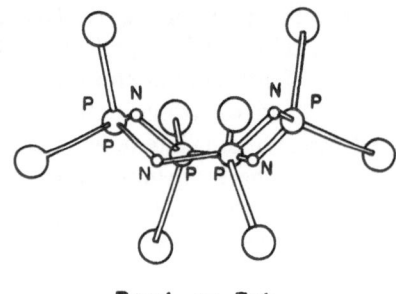

Bond length and angle variation in geminal and spirocyclic derivatives.

These bond length variations have been rationalized in terms of different degrees of ring π-bonding within a molecule. Strong electronegative substituents contract the d-orbital at the phosphorus atom and make them less available for

π-bonding with ring nitrogens. In other words, electron releasing substituents fill the vacant d orbitals at phosphorus and push the nitrogen lone pairs towards PCl_2 groups and increase P–N bond order in the $N-PCl_2-N$ segment.

In the geminal disubstituted cyclophosphazenes the ring bond angles are in the following order: $\alpha < \beta$ (at phosphorus) and $\gamma > \delta$ (at nitrogen). A reverse in this order, $\beta < \alpha$ and $\delta > \gamma$ is observed for the geminal tetrasubstituted cyclo-triphosphazenes. Similar bond angle variations have been observed for mono-spiro and dispiro derivatives resulting from difunctional reagents (Table 8). Examples for disubstituted geminal and spiro derivatives are given in Figs. 11 and 12.

To generalize, electron donating groups on phosphorus tend to:

1) reduce the endocyclic phosphorus angle with the increase in the exocyclic phosphorus angle.
2) weaken the P–N bonds in the $N-PR_2-N$ segment. This decrease in P–N bond length is more pronounced when phosphorus is directly linked to a metal which facilitates exocyclic P–M π-bond formation (cf. entries 30 and 31 in Table 7).
3) increase the bond angles at the succeeding nitrogens on both sides.
4) decrease the bond angle at the nitrogen flanked by phosphorus containing relatively weak electron donating groups i.e., PCl_2 or PClR.

Protonation of the ring nitrogen also reveals interesting bond length variations. The P–N bond lengths involving the protonated nitrogen atom (in P–NH–P segment) are longer than in the free base. The exocyclic P–N bond lengths on the phosphorus atoms adjacent to the protonation site are shorter than in the free base. Thus, for $[gem-N_3P_3(NHPr^i)_4Cl_2H]^+Cl^-$ the average P–N bond length at the P–NH–P segment is 1.665(5) Å which is considerably longer than the corresponding bond lengths in the free base [1.589(3) Å], $N_3P_3(NHPr^i)_4Cl_2$ [12]. The exocyclic P–N bond length in the free base is

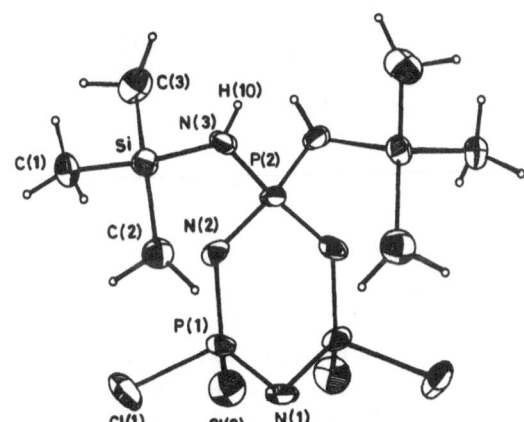

Fig. 11. Crystal structure of $N_3P_3Cl_4$ $(NHSiMe_3)_2$ (from Ref. 298)

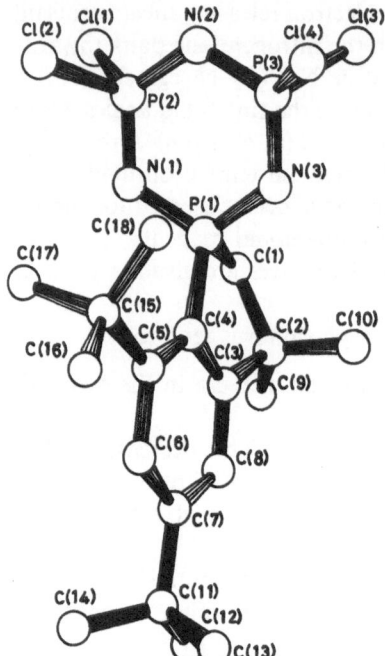

Fig. 12. Molecular structure of a novel spirocyclic compound. (from Ref. 357)

1.642(4) Å while in the HCl adduct it is shortened to 1.609(5) Å. Similar effects have also been realized when the ring nitrogen coordinates to the metal. These are atributed to the decreased π-bonding between the protonated or metal coordinated nitrogen and the adjacent phosphorus atoms because of the non-availability of nitrogen lone pair electrons and a increased π-bonding between the phosphorus and exocyclic nitrogens.

Attempts at correlating endocyclic phosphorus bond angles with the phosphorus chemical shifts and P–N bond distances with [35]Cl NQR frequencies have been made [314]. It appears that for closely related compounds there exists a narrow relationship between ^{31}P chemical shifts and endocyclic phosphorus angles [96, 358]. No generalized trend has yet emerged. The differences in the P–Cl bond lengths in PCl_2 and $PRCl$ units have been correlated with the [35]Cl NQR frequencies [314].

11 Summary and Conclusions

Cyclophosphazenes have continued to attract a lot of attention in the last decade. In a significant development the first four membered cyclophosphazene has been synthesised and structurally characterised (Sect. 2). Studies on the reactivity of $N_3P_3Cl_6$ and $N_4P_4Cl_8$ with amines continues to be an attractive area. Quantitative kinetic data have been obtained for the reactions of certain amines with $N_3P_3Cl_6$. This has helped in understanding the factors responsible for the stereo and regio selectivity observed in these reactions (Sect. 3). However, such data are limited in scope with reactions involving hydroxy, thiolato, and many other nucleophilic reagents. In the reactions of $N_4P_4Cl_8$ with amines the formation of "bicyclic" products has been observed under certain conditions. The mechanism of this reaction is now reasonably well understood (Sect. 4).

A number of studies have been carried out, primarily on $N_3P_3Cl_6$ with several types of difunctional reagents. These reactions are complex and depending on the nature of the reagent a product preference is seen (Sect. 5). The last decade has also seen the extension of the nucleophilic substitution reactions at phosphorus to several versatile reagents including main group and transition metal organometallic compounds. A rich chemistry has ensued, and is continuing to develop in this area (Sect. 7). A significant outcome of these pursuits is the isolation of the versatile hydridocyclophosphazenes whose chemistry has a vast scope in view of the reactivity of the P–H bond as well as the tautomerism (H migration from phosphorus to nitrogen) observed for these derivatives (Sect. 8). Several interesting rearrangements and regio and stereochemical transformations have been observed. However, the mechanisms of some of these transformations are not yet fully understood (Sect. 6).

The coordination chemistry of cyclophosphazenes has been expanding vigorously with synthesis of several derivatives that can function as ligands through endocyclic nitrogen atoms and/or exocyclic ligating groups. Many metal clusters involving cyclophosphazenes as templates have been prepared. Clearly several new developments are likely in this area in the coming years (Sect. 9).

The solution and solid state structural characterisation of cyclophosphazenes has been carried out mainly by the use of ^{31}P and ^{1}H NMR and single crystal X-ray methods. Although there is a huge wealth of data concerning the phosphorus chemical shifts and coupling constants, there is still no satisfactory theoretical rationalization that can draw correlations between chemical shifts and structural features. However, empirical correlations continue to be successfully used (Sect. 10). The variations in bond lengths and bond angles as determined in the solid state mainly by X-ray structural methods have been accounted for by using existing theories of bonding in cyclophosphazenes. Although a precise picture of bonding has not yet emerged in cyclophosphazenes, most of the bond length and bond angle variations observed in different cyclophosphazene derivatives are reasonably explained by invoking the different types of multiple bonding possibilities in these compounds (Sect. 10.4).

The versatile and rich chemistry that is exhibited by these P–N ring systems and their exciting structural features suggest that cyclophosphazenes will continue to attract a lot of attention in the coming years.

Acknowledgements. We are very much grateful to DST (India) for financial assistance.

12 References

1. Allen CW (1987) In: Haiduc I, Sowerby DB (eds) The chemistry of inorganic homo and heterocycles, Academic, vol 2, p 501
2. Allen CW (1991) Chem Rev 91: 133
3. Krishnamurthy SS, Sau AC, Woods M (1978) Adv Inorg Chem Radiochem 21: 41
4. Potin Ph, De Jaeger R (1991) Eur Polym J 27: 341
5. Singler RE, Sennett MS, Willingham RA (1987) Am Chem Symp Ser 360: 268
6. Allen CW (1987) Am Chem Symp Ser 360: 290
7. Chandrasekhar V, Justin Thomas KR (1992) Appl Organomet Chem in press
8. Allcock HR, Desorcie JL, Riding GH (1987) Polyhedron 6: 119
9. Paddock NL (1986) Int Rev Phys Chem 5: 161
10. Haddon RC (1985) Chem Phys Lett 120: 372
11. Ferris K, Friedman P, Friedrich DM (1988) Int J Quan Chem Quant Chem Symp 22: 207
12. Shaw RA (1976) Z Naturforsch 31b: 641
13. Allcock HR (1971) Phosphorus-nitrogen compounds, Academic, New York
14. Allen CW (1986–1991) In: Organophosphorus chemistry, Royal Society of Chemistry, London, vols 16–22
15. Emsley J, Udy PB (1970) J Chem Soc A 3025
16. Emsley J, Udy PB (1971) J Chem Soc A 768
17. Mao TJ, Dresdner RD, Young MA (1959) J Am Chem Soc 81: 1020
18. Tasaka A, Glemser O (1974) Z Anorg Allg Chem 409: 163
19. Roesky HW (1972) Angew Chem Int Ed Engl 11: 642
20. Emsley J, Paddock NL (1968) J Chem Soc A 2590
21. Green B, Sowerby DB, Clare PC (1971) J Chem Soc A 3487
22. Paddock NL, Patmore DJ (1976) J Chem Soc Dalton Trans 1029
23. Neilson RH, Wisian-Neilson P (1988) Chem Rev 88: 541
24. Wisian-Neilson P, Neilson RH (1980) J Am Chem Soc 102: 2848
25. Allcock HR (1972) Chem Rev 72: 315
26. Baceiredo A, Bertrand G, Majoral JP, Sicard G, Juad J, Galy J (1984) J Am Chem Soc 106: 6088
27. Granier M, Baceiredo A, Dartiguenave Y, Dartiguenave M, Menu M-J, Bertrand G (1990) J Am Chem Soc 112: 6277
28. Shaw RA (1980) Pure and Appl Chem 52: 1063
29. Van der Huizen AA, Van de Grampel JC, Rusch JW, Witting T, Bolhuis F, Meetsma A (1986) J Chem Soc Dalton Trans 1317
30. Van der Huizen AA, Jakel AP, Rusch JW, Van de Grampel JC (1981) Recl Trav Chim Pays-Bas 100: 343
31. Kobayashi Y, Chasin LA, Clapp LB (1963) Inorg Chem 2: 212
32. Ottmann GF, Agahigian H, Hooks H, Vickers GD, Kober EH, Rätz RFW (1964) Inorg Chem 3: 753
33. Rätz RFW, Kober EH, Grundmann C, Ottmann GF (1964) Inorg Chem 3: 757
34. Fiestel GR, Moeller TJ (1967) J Inorg Nucl Chem 29: 2731
35. (a) Brian Z, Goldschmidt JME (1978) Synth React Inorg Met.-Org Chem 8: 185 (b) Biddlestone M, Shaw RA (1973) J Chem Soc Dalton Trans 2740
36. Lehr WZ (1967) Z Anorg Allg Chem 352: 27
37. Das RN, Shaw RA, Smith BC, Woods M (1973) J Chem Soc Dalton Trans 709
38. Desai VB, Shaw RA, Smith BC (1970) J Chem Soc A 2023
39. Ganapathiappan S, Krishnamurthy SS (1987) J Chem Soc Dalton Trans 579
40. Lingley DJ, Shaw RA, Yu HS (1980) Inorg Nucl Chem Letts 16: 219
41. Keat R, Shaw RA (1965) J Chem Soc 2215
42. Green B, Sowerby DB (1971) J Inorg Nucl Chem 33: 3687
43. Goldschmidt JME, Sadeh U (1980) J Inorg Nucl Chem 42: 618
44. Keat R, Shaw RA (1966) J Chem Soc A 908
45. Krishnamurthy SS, Rao MNS, Vasudeva Murthy AR, Shaw RA, Woods M (1976) Ind J Chem 14a: 823
46. Brian Z, Goldschmidt JME (1979) J Chem Soc Dalton Trans 1017
47. Ray SK, Shaw RA (1961) J Chem Soc 872

48. Lingley DJ, Shaw RA, Woods M, Krishnamurthy SS (1978) Phosphorus Sulfur 4: 379
49. Das SK, Keat R, Shaw RA, Smith BC (1965) J Chem Soc 5032
50. Allen CW, Mackay JA (1986) Inorg Chem 25: 4628
51. Smardijk AA, De Ruiter B, Van der Huizen AA, Van de Grampel JC (1982) Recl Trav Chim Pays-Bas 101: 270
52. (a) Nabi SN, Shaw RA, Stratton C (1969) Chem Ind (London) 166 (b) Hasa M, Shaw RA, Woods M (1975) J Chem Soc Dalton Trans 2202
53. Keat R, Shaw RA (1968) Angew Chem Int Ed Engl 7: 212
54. Goldschmidt JME, Licht E (1979) J Chem Soc Dalton Trans 1012
55. Goldschmidt JME, Licht E (1981) J Chem Soc Dalton Trans 107
56. Goldschmidt JME, Goldstein R (1981) J Chem Soc Dalton Trans 1283
57. Krishnamurthy SS, Sundaram PM (1982) J Chem Soc Dalton Trans 67
58. Katti KV, Krishnamurthy SS (1985) J Chem Soc Dalton Trans 285
59. Ganapathiappan S, Krishnamurthy SS (1987) J Chem Soc Dalton Trans 585
60. Goldschmidt JME, Licht E (1972) J Chem Soc Dalton Trans 728, 732
61. Krishnamurthy SS, Sau AC, Vasudeva Murthy AR, Keat R, Shaw RA, Woods M (1977) J Chem Soc Dalton Trans 1980
62. Krishnamurthy SS, Ramachandran K, Woods M (1981) Phosphorus Sulfur 9: 323
63. Krishnamurthy SS, Rao MNS, Vasudeva Murthy AR, Shaw RA, Woods M (1978) Inorg Chem 17: 1527
64. Millington D, Sowerby DB (1972) J Chem Soc Dalton Trans 2035
65. Krishnamurthy SS, Sau AC, Vasudeva Murthy AR, Keat R, Shaw RA, Woods M (1976) J Chem Soc Dalton Trans 1405
66. Krishnamurthy SS (1989) Phosphorus Sulfur Silicon 41: 375
67. Krishnamurthy SS, Sundaram PM, Woods M (1982) Inorg Chem 21: 406
68. Narayana Swamy PY, Dhathathreyan KS, Krishnamurthy SS (1985) Inorg Chem 24: 640
69. Contractor SR, Kilic Z, Shaw RA (1987) J Chem Soc Dalton Trans 2023
70. Krishnamurthy SS, Sau AC, Vasudeva Murthy AR, Shaw RA, Woods M, Keat R (1977) J Chem Res(S) 70
71. Krishnamurthy SS, Ramachandran K, Sau AC, Shaw RA, Vasudeva Murthy AR, Woods M (1979) Inorg Chem 18: 2010
72. Finer EG, Harris RK, Bond MR, Keat R, Shaw RA (1970) J Mol Spect 33: 72
73. (a) Schmulz JL, Allcock HR (1975) Inorg Chem 14: 2433 (b) Allcock HR, Ngo DC, Parvez M, Visscher K (1992) J Chem Soc Dalton Trans 1687
74. Zeleneva TP, Antonov IV, Stepanov BI (1973) J Gen Chem USSR 43: 1000
75. Dhathathreyan KS, Krishnamurthy SS, Woods M (1982) J Chem Soc Dalton Trans 2151
76. Kumara Swamy KC, Krishnamurthy SS, Vasudeva Murthy AR, Shaw RA, Woods M (1986) Ind J Chem 25a: 1004
77. Karthikeyan S, Krishnamurthy SS, Woods M (1984) Z Anorg Allg Chem 513: 231
78. Dell D, Fitzsimmons BW, Keat R, Shaw RA (1966) J Chem Soc A 1680
79. Allcock HR, Dembek AA, Mang MN, Riding GH, Parvez M, Visscher K (1992) Inorg Chem 31: 2734
80. (a) Thomas B, Schadow H, Shceeler H (1975) Z Chem 15: 26 (b) Thomas B, Grossman G (1979) Z Anorg Allg Chem 448: 100 (c) Thomas B, Grossman G (1979) Z Anorg Allg Chem 448: 107
81. (a) Neicke E, Glemsor O, Roesky HW (1969) Z Naturforsch 24b: 1187 (b) Carrol AP, Shaw RA, Woods M (1973) J Chem Soc Dalton Trans 2736
82. Chandrasekhar V, Muralidhara MG, Selvaraj II (1990) Heterocycles 31: 2231
83. Shaw RA (1989) Phosphorus Sulfur Silicon 45: 103
84. Labarre J-F (1985) Top Curr Chem 129: 173
85. Alkubaisi AH, Deutsch WF, Hursthouse MB, Parkes HG, Shaw LS (neé Gözen), Shaw RA (1986) Phosphorus Sulfur 28: 229
86. Krishnamurthy SS, Ramachandran K, Vasudeva Murthy AR, Shaw RA, Woods M (1980) J Chem Soc Dalton Trans 840
87. Chivers T, Hedgeland R (1972) Can J Chem 50: 1017
88. Ruiter BD, Kuipers G, Bijlaart JH, Van de Grampel JC (1982) Z Naturforsch 37b: 1425
89. Chandrasekhar V, Krishnamurthy SS, Vasudeva Murthy AR, Shaw RA, Woods M (1981) Inorg Nucl Chem Letts 17: 181
90. Chandrasekhar V, Krishnamurthy SS, Manohar H, Vasudeva Murthy AR, Shaw RA, Woods M (1984) J Chem Soc Dalton Trans 621

91. Harris PJ, Williams KB (1984) Inorg Chem 23: 1496
92. Deutch WF, Shaw RA (1990) Phosphorus Sulfur Silicon 47: 119
93. Deutch WF, Shaw RA (1988) J Chem Soc Dalton Trans 1757
94. Chandrasekhar V, Karthikeyan S, Krishnamurthy SS, Woods M (1985) Ind J Chem 24a: 379
95. Alkubaisi AH, Shaw RA (1989) Phosphorus Sulfur Silicon 45: 7
96. Contractor SR, Hursthouse MB, Shaw LS (neé Gözen), Shaw RA, Yilmaz H (1985) Acta Cryst 41b: 122
97. Alkubaisi AH, Al-madfa HA, Deutsch WF, Hursthouse MB, Parkes HG, Shaw LS (neé Gözen), Shaw RA (1986) Phosphorus Sulfur 28: 277
98. Contractor SR, Hursthouse MB, Parkes HG, Shaw LS (neé Gözen), Shaw RA, Yilmaz H (1984) J Chem Soc Chem Commun 675
99. Contractor SR, Hursthouse MB, Parkes HG, Shaw LS (neé Gözen), Shaw RA, Yilmaz H (1986) Phosphorus Sulfur 28: 267
100. Alkubaisi AH, Shaw RA (1991) Phosphorus Sulfur Silicon 55: 49
101. Guerch G, Graffeuil M, Lebarre J-F, Enjalbert R, Lahana R, Sournies F (1982) J Mol Struct 95: 237
102. Enjalbert R, Guerch G, Labarre J-F, Galy J (1982) Z Krist 160: 249
103. El Murr N, Lahana R, Labarre J-F, Declercq J-P (1984) J Mol Struct 117: 73
104. Guerch G, Labarre J-F, Roques R, Sournies F (1982) J Mol Struct 96: 113
105. Guerch G, Labarre J-F, Lahana R, Roques R, Sournies F (1983) J Mol Struct 99: 275
106. Castera P, Faucher J-P, Guerch G, Lahana R, Mahmoun A, Sournies F, Labarre J-F (1985) Inorg Chim Acta 108: 29
107. Castera P, Faucher J-P, Granier M, Labarre J-F (1987) Phosphorus Sulfur 32: 37
108. Labarre J-F, Guerch G, Sournies F, Lahana R, Enjalbert R, Galy J (1984) J Mol Struct 116: 75
109. Sournies F, Lahana R, Labarre J-F (1985) Inorg Chim Acta 101: 31
110. Sournies F, Labarre J-F, Spreafico F, Fillippeschi S, Xing Quan Jin (1986) J Mol Struct 147: 161
111. Castera P, Faucher J-P, Labarre J-F, Perly B (1987) J Mol Struct 160: 365
112. Enjalbert R, Galy J, Castera P, Labarre J-F (1988) Acta Cryst 44c: 1813
113. Sournies F. Bakili AE, Labarre J-F, Perly B (1989) J Mol Struct 196: 201
114. Enjalbert R, Galy J, Bakili AE, Castera P, Faucher J-P, Sournies F, Labarre J-F (1989) J Mol Struct 195: 207
115. Bakili AE, Castera P, Faucher J-P, Sournies F, Labarre J-F (1989) J Mol Struct 195: 21
116. Sournies F, Labarre J-F, Harris PJ, Williams KB (1984) Inorg Chim Acta 90: L61
117. Enjalbert R, Galy J, Sournies F, Labarre J-F (1990) J Mol Struct 221: 253
118. Sassus J-L, Graffeuil M, Castera P, Labarre J-F (1985) Inorg Chim Acta 108: 23
119. Guerch G, Bonnett J-P, Labarre J-F (1989) J Mol Struct 196: 221
120. Guerch G, Labarre J-F (1989) J Mol Struct 195: 11
121. Zanin B, Faucher J-P, Labarre J-F (1990) Inorg Chim Acta 172: 147
122. Cameron TS, Linden A, Bakili AE, Castera P, Faucher J-P, Graffeuil M, Sournies F, Labarre J-F (1989) J Mol Struct 212: 281
123. Sournies F, Bakili AE, Zanin B, Labarre J-F, Juad J (1990) J Mol Struct 220: 63
124. Zanin B, Sournies F, Labarre J-F, Enjalbert R, Galy J (1990) J Mol Struct 240: 77
125. Sournies F, Castera P, Bakili AE, Labarre J-F (1990) J Mol Struct 221: 239
126. Sournies F, Castera P, Faucher J-P, Graffeuil M, Labarre J-F (1990) J Mol Struct 221: 245
127. Perly B, Berthault P, Bonneth J-P, Labarre J-F (1988) J Mol Struct 176: 285
128. CAstera P, Faucher J-P, Graffeuil M, Labarre J-F (1988) J Mol Struct 176: 295
129. Veith M, Kross M, Labarre J-F (1991) J Mol Struct 243: 189
130. Enjalbert R, Galy J, Zanin B, Bonnett J-P, Sournies F, Labarre J-F (1991) J Mol Struct 246: 123
131. Allcock HR, Siegel LA (1964) J Am Chem Soc 86: 5140
132. Allcock HR, Stein MT, Stankoja (1971) J Am Chem Soc 93: 3173
133. Allcock HR, Kugel RL (1969) J Am Chem Soc 91: 5452
134. Allcock HR, Kugel RL, Moore GY (1975) Inorg Chem 14: 283
135. Brandt K, Jedlinski Z (1980) J Org Chem 45: 1672
136. Allcock HR, Siegel LA (1964) J Am Chem Soc 86: 5140
137. Allcock HR, Levin ML (1985) Macromolecules 18: 1324
138. Al-Madfa, Homaid A, Shaw RA (1991) Phosphorus Sulfur Silicon 55: 59
139. Alkubaisi AH, Parkes HG, Shaw RA (1989) Heterocycles 28: 347

140. Kumara Swamy KC, Krishnamurthy SS (1984) Ind J Chem 23a: 717
141. Keat R, Shaw RA (1964) Chem and Ind 1232
142. Keat R, Shaw RA, Stratton C (1965) J Chem Soc 2223
143. Keat R, Shaw RA (1965) J Chem Soc 4067
144. Shaw RA (1988) J Organomet Chem 341: 357
145. Friedman N, Goldschmidt JME, Sadeh U, Segev M (1981) J Chem Soc Dalton Trans 103
146. Allcock HR, Desorcie JL, Wagner LJ (1985) Inorg Chem 24: 333
147. Fieldhouse JW, Graves DF (1981) Am Chem Symp Ser 171: 315
148. Kobayashi E (1976) Bull Chem Soc Jpn 49: 3524
149. Allcock HR (1979) Acc Chem Res 12: 351
150. Parvez M, Kwon S, Allcock HR (1991) Acta Cryst 47c: 466
151. Dhathathreyan KS, Krishnamurthy SS, Vasudeva Murthy AR, Cameron TS, Chan C, Shaw
     RA, Woods M (1980) J Chem Soc Chem Commun 231
152. Bullen GJ, Dann PE, Evans ML, Hursthouse MB, Shaw RA, Wait K, Woods M, Yu HS (1976)
     Z Naturforsch 31b: 995
153. Dhathathreyan KS, Krishnamurthy SS, Vasudeva Murthy AR, Shaw RA, Woods M (1982)
     J Chem Soc Dalton Trans 1549
154. De Ruiter B, Winter H, Wilting T, Van de Grampel JC (1984) J Chem Soc Dalton Trans 1027
155. Fitzsimmons BW, Hewlett C, Shaw RA (1964) J Chem Soc 4459
156. Fitzsimmons BW, Shaw RA (1961) Proc Chem Soc 258
157. Ansell GB, Bullen GJ (1968) J Chem Soc A 3026
158. Bullen GJ, Paddock NL, Patmore DJ (1977) Acta Cryst 33b: 1367
159. Bullen GJ, Williams SJ, Paddock NL, Patmore DJ (1981) Acta Cryst 37b: 607
160. Dhathathreyan KS, Krishnamurthy SS, Vasudeva Murthy AR, Shaw RA, Woods M (1979)
     Inorg Nucl Chem Letts 15: 109
161. Dhathathreyan KS, Krishnamurthy SS, Vasudeva Murthy AR, Shaw RA, Woods M (1981)
     J Chem Soc Dalton Trans 1928
162. Fitzsimmons BW, Hewlett C, Shaw RA (1965) J Chem Soc 7432
163. Mochel VD, Cheng TC (1978) Macromolecules 11: 176
164. Cheng TC, Mochel VD, Adams HE, Longo TF (1980) Macromolecules 13: 158
165. Karthikeyan S, Vyas K, Krishnamurthy SS, Kameron TS, Vincent BR (1988) J Chem Soc
     Dalton Trans 1371
166. Ferrar WT, Distefano FV, Allcock HR (1980) Macromolecules 13: 1345
167. Karthikeyan S, Krishnamurthy SS (1991) J Chem Soc Dalton Trans 299
168. Fincham JK, Hursthouse MB, Parkes HG, Shaw LS (neé Gözen), Shaw RA (1985) J Chem Soc
     Chem Commun 252
169. Fincham JK, Hursthouse MB, Parkes HG, Shaw LS (neé Gözen), Shaw RA (1986) Phosphorus
     Sulfur 28: 185
170. Fincham JK, Parkes HG, Shaw LS (neé Gözen), Shaw RA, Hursthouse MB (1988) J Chem Soc
     Dalton Trans 1169
171. Calhoun HP, Oakley RT, Paddock NL (1975) J Chem Soc Chem Commun 454
172. Oakley RT, Paddock NL (1977) Can J Chem 55: 3651
173. Oakley RT, Paddock NL, Rettig SJ, Trotter J (1977) Can J Chem 55: 2534
174. Paddock NL, Ranganathan TN, Todd SM (1971) Can J Chem 49: 104
175. Ranganathan TN, Todd SM, Paddock NL (1973) Inorg Chem 12: 316
176. Dupont JG, Allen CW (1978) Inorg Chem 17: 3093
177. Allen CW, Desorcie JL, Ramachandran K (1984) J Chem Soc Dalton Trans 2843
178. Chivers T (1971) Inorg Nucl Chem Letts 7: 827
179. Allen CW, Bright RP (1983) Inorg Chem 22: 1291
180. Allen CW, Moeller T (1968) Inorg Chem 7: 2177
181. Ramachandran K, Allen CW (1982) J Am Chem Soc 104: 2396
182. Allen CW, Toch PL (1981) Inorg Chem 20: 8
183. Shaw JC, Allen CW (1986) Inorg Chem 25: 4632
184. Allen CW, White AJ (1974) Inorg Chem 13: 1220
185. Searle HT, Dyson J, Ranganathan TN, Paddock NL (1975) J Chem Soc Dalton Trans 203
186. Allen CW, Green JC (1980) Inorg Chem 19: 1719
187. Allen CW (1977) J Organomet Chem 125: 215
188. Allcock HR, Harris PJ, Nissan RA (1981) J Am Chem Soc 103: 2256
189. Biddlestone M, Shaw RA (1973) Phosphorus 3: 95

190. Harris PJ, Fadeley CL (1983) Inorg Chem 22: 561
191. Winter H, van de Grampel JC (1986) J Chem Soc Dalton Trans 1269
192. Winter H, van de Grampel JC (1984) J Chem Soc Chem Commun 489
193. Winter H, van de Grampel JC (1984) Recl Trav Chim Pays-Bas 103: 241
194. Scopelianos AG, O'Brien JP, Allcock HR (1980) J Chem Soc Chem Commun 198
195. Allcock HR, Scopelianos AG, O'Brien JP, Bernheim MY (1981) J Am Chem Soc 103: 350
196. Allcock HR, Lavin KD, Riding GH, Suszko PR, Whittle RR (1984) J Am Soc 106: 2337
197. Riding GH, Parvez M, Allcock HR (1986) Organometallics 5: 2153
198. Allcock HR, Lavin KD, Riding GH, Whittle RR, Parvez M (1986) Organometallics 5: 1626
199. Suszko PR, Whittle RR, Allcock HR (1982) J Chem Soc Chem Commun 960
200. Lavin KD, Riding GH, Parvez M, Allcock HR (1986) J Chem Soc Chem Commun 117
201. Allcock HR, Fuller TJ, Evans TL (1980) Macromolecules 13: 1325
202. Allcock HR, Lavin KD, Riding GH, Whittle RR (1984) Organometallics 3: 663
203. Allcock HR, Lavin KD, Riding GH (1985) Macromolecules 18: 1340
204. Allcock HR, Connolly MS, Whittle RR (1983) Organometallics 2: 1514
205. Harris PJ, Desorcie JL, Allcock HR (1981) J Chem Soc Chem Commun 852
206. Allcock HR, Desorcie JL, Harris PJ (1983) J Am Chem Soc 105: 2814
207. Harris PJ, Williams KB, Fisher BL (1984) J Org Chem 49: 406
208. Allcock HR, Brennan DJ, Graaskamp JM, Parvez M (1986) Organometallics 5: 2434
209. Biddlestone M, Shaw RA (1969) J Chem Soc A 178
210. Biddlestone M, Shaw RA (1971) J Chem Soc A 2715
211. Biddlestone M, Shaw RA (1965) J Chem Soc Chem Commun 205
212. Biddlestone M, Shaw RA (1970) J Chem Soc A 1750
213. Allen CW, Dieck RL, Brown P, Moeller T, Schmulback CD, Cook AG (1978) J Chem Soc Dalton Trans 173
214. Feidt MK, Moeller T (1968) J Inorg Nucl Chem 30: 2351
215. Tesai G, Slota PJ (1960) Proc Chem Soc London 404
216. Harris PJ, Allcock HR (1978) J Am Chem Soc 100: 6512
217. Allcock HR, Harris PJ (1979) J Am Chem Soc 101: 6221
218. Allcock HR, Harris PJ (1981) Inorg Chem 20: 2844
219. Harris PJ, Allcock HR (1979) J Chem Soc Chem Commun 714
220. Allcock HR, Harris PJ, Connolly MS (1981) Inorg Chem 20: 11
221. Allcock HR, Harris PJ, Nissan RA (1981) J Am Chem Soc 103: 2256
222. Harris PJ, Schwalke MA, Liu U, Fischer BL (1983) Inorg Chem 22: 1812
223. Buwalda PL, Steenbergen A, Oosting GE, Van de Grampel JC (1990) Inorg Chem 29: 2658
224. Jackson LA, Harris PJ (1988) Inorg Chem 27: 4338
225. Harris PJ, Jackson LA (1983) Organometallics 2: 1477
226. Calhoun HP, Lindstrom RH, Oakley RT, Paddock NL, Todd SM (1975) J Chem Soc Chem Commun 343
227. Gallicano KD, Oakley RT, Paddock NL, Sharma RD (1981) Can J Chem 59: 2654
228. Allcock HR, Suszko PR, Evans, TL (1982) Organometallics 1: 1443
229. McBee ET, Okuhara K, Morton CJ (1965) Inorg Chem 4: 1672
230. Das S, Shaw RA, Smith BC (1973) J Chem Soc Dalton Trans 1883
231. Das S, Shaw RA, Smith BC (1974) J Chem Soc Dalton Trans 1610
232. Das SK, Hasan M, Shaw RA, Smith BC, Woods M (1979) Z Naturforsch 34b: 58
233. Biddlestone M, Shaw RA (1973) J Chem Soc Dalton Trans 2740
234. Allen CW, Tsang FY, Moeller T (1968) Inorg Chem 7: 2177
235. Allen CW, Brunst GE, Perlman ME (1980) Inorg Chim Acta 41: 265
236. Allen CW, Bedell S, Pennington WT, Cordes AW (1985) Inorg Chem 24: 1653
237. Schmidpeter A, Ebeling J, Stary H, Weingand C (1972) Z Anorg Allg Chem 394: 171
238. Bermann M, Vanwazer JR (1972) Inorg Chem 11: 209
239. Schmidpeter A, Blanck K, Smetana H, Weingand C (1977) Synth React Inorg Met-Org Chem 7: 1
240. Schmidpeter A, Högel J (1978) Chem Ber 111: 3867
241. Schmidpeter A, Högel J, Ahmed FR (1976) Chem Ber 109: 1911
242. Schmidpeter A, Blanck K, Ahmed FR (1976) Angew Chem Int Ed Engl 15: 488
243. Schmidpeter A, Rossnecht H (1974) Chem Ber 107: 3146
244. Schmidpeter A, Elietz H (1975) Chem Ber 108: 1454
245. Högel J, Schmidpeter A (1979) Z Anorg Allg Chem 458: 168

246. Schmidpeter A, Blanck K, Hess H, Riffel H (1980) Angew Chem Int Ed Engl 19: 650
247. Dash KC, Schmidpeter A, Schmidbaur H (1980) Z Naturforsch 35b: 1286
248. Cotton FA, Rusholme GA, Shaver A (1973) J Coord Chem 99
249. Coxon GE, Sowerby DB (1969) J Chem Soc A 3012
250. Allcock HR, Bissel, EC, Shawl LJ (1973) Inorg Chem 12: 2963
251. Macdonald AL, Trotter J (1974) Can J Chem 52: 734
252. Trotter J, Whitlow SH (1970) J Chem Soc A 455
253. Allcock HR, Allen RW, O'Brien JP (1977) J Am Chem Soc 99: 3954
254. O'Brien JP, Allen RW, Allcock HR (1979) Inorg Chem 18: 2230
255. Allen RW, O'Brien JP, Allcock HR (1977) J Am Chem Soc 99: 3987
256. Trotter J, Whitlow SH (1970) J Chem Soc A 460
257. Calhoun HP, Trotter J (1974) J Chem Soc Dalton Trans 382
258. Paddock NL, Ranganathan TN, Wingfield JN (1972) J Chem Soc Dalton Trans 1578
259. Paddock NL, Ranganathan TN, Rettig SJ, Sharma R, Trotter J (1981) Can J Chem 59: 2429
260. Calhoun HP, Paddock NL, Wingfield JN (1975) Can J Chem 53: 1765
261. March WC, Trotter J (1971) J Chem Soc A 1482
262. Harrison W, Trotter J (1979) J Chem Soc Dalton Trans 61
263. Gallicano KD, Paddock NL, Rettig SJ, Trotter J (1981) Can J Chem 59: 2435
264. Gallicano KD, Paddock NL (1982) Can J Chem 60: 521
265. Bertani R, Facchin G, Gleria M (1989) Inorg Chim Acta 165: 73
266. Allcock HR, Manners I, Mang MN, Parvez M (1990) Inorg Chem. 29: 522
267. Allcock HR, Lavin KD, Tollefson NM, Evans TL (1983) Organometallics 2: 267
268. Allcock HR, Neenan TX (1986) Macromolecules 19: 1495
269. Selvaraj II, Chandrasekhar V, Reddy D, Chandrashekar TK (1991) Heterocycles 32: 701
270. Allcock HR, Greigger PP, Gardner JE, Schmultz JL (1979) J Am Chem Soc 101: 606
271. Allcock HR, Neenan TX, Boso B (1985) Inorg Chem 24: 2656
272. Allcock HR, Nissan RA, Harris PJ, Whittle RR (1984) Organometallics 3: 432
273. Allcock HR, Scopelianos AG, Whittle RR, Tollefson NM (1983) J Am Chem Soc 105: 1316
274. Calhoun HP, Paddock NL, Trotter J (1973) J Chem Soc Dalton Tans 2708
275. Galicano KD, Paddock NL, Rettig SJ, Trotter J (1979) Inorg Nucl Chem Letts 15: 417
276. Chandrasekharan A, Krishnamurthy SS, Nethaji M (1991) Curr Sci 60: 700
277. Justin Thomas KR, Chandrasekhar V, Pal P, Scott SS, Hallford R, Cordes AW (1992) Inorg Chem (in press)
278. Allcock HR, Dembek AA, Bennett JC, Manners I, Parvez M (1991) Organometallics 10: 1865
279. Allcock HR, Greigger PP (1979) J Am Chem Soc 101: 2492
280. Allcock HR, Greigger PP, Wagner LJ, Bernheim MY (1981) Inorg Chem 20: 716
281. Allcock HR, Wagner LJ, Levin ML (1983) J Am Chem Soc 105: 1321
282. Allcock HR, Suszko PR, Wagner LJ, Whittle RR, Boso B (1984) J Am Chem Soc 106: 4966
283. Suszko PR, Whittle RR, Allcock HR (1982) J Chem Soc Chem Commun 649
284. Allcock HR, Suszko PR, Wagner LJ, Whittle RR, Boso B (1985) Organometallics 4: 446
285. Allcock HR, Riding GH, Whittle RR (1984) J Am Chem Soc 106: 5561
286. Nissan RA, Connolly MS, Mirabelli MGL, Whittle RR, Allcock HR (1983) J Chem Soc Chem Commun 822
287. Allcock HR, Mang MN, Mc Donnell GS, Parvez M (1987) Macromolecules 20: 2060
288. Krishnamurthy SS, Woods M (1987) Annu Rep NMR Spect 19: 1
289. Kumara Swamy KC, Krishnamurthy SS (1986) Inorg Chem 25: 920
290. Kumara Swamy KC, Krishnamurthy SS (1983) Phosphorus Sulfur 18: 241
291. Krishnamurthy SS, Ramabrahmam P, Vasudeva Murthy AR, Shaw RA, Woods M (1985) Z Anorg Allg Chem 522: 226
292. Kumara Swamy KC, Poojari MD, Krishnamurthy SS, Manohar H (1984) Z Naturforsch 39b: 615
293. Kumara Swamy KC, Poojari MD, Krishnamurthy SS, Manohar H (1985) J Chem Soc Dalton Trans 1581
294. Krishnamurthy SS, Ramabrahmam P, Vasudeva Murthy AR, Shaw RA, Woods M (1980) Inorg Nucl Chem Letts 16: 215
295. Krishnamurthy SS, Ramachandran K, Vasudeva Murthy AR, Shaw RA, Woods M (1977) Inorg Nucl Chem Letts 13: 407
296. Keat R, Shaw RA, Woods M (1976) J Chem Soc Dalton Trans 1582
297. Lensink C, De Ruiter B, Van de Grampel JC (1984) J Chem Soc Dalton Trans 1521

298. Allen CW, Brown DE, Cordes AW, Craig SC (1988) J Chem Soc Dalton Trans 1405
299. Ramachandran K, Allen CW (1983) Inorg Chem 22: 1445
300. Brown DE, Allen CW (1987) Inorg Chem 26: 934
301. Reuben J (1987) Magn Reson Chem 25: 1049
302. Finer EG, Harris RK (1971) Prog Nucl Magn Reson Spect 6: 61
303. Schumann K, Schmidpeter A (1973) Phosphorus 3: 51
304. Fincham JK, Hursthouse MB, Parkes HG, Shaw LS (neé Gözen), Shaw RA (1986) Acta Cryst 42b: 462
305. Paddock NL (1964) Quart Rev Chem Soc 18: 168
306. Moeller T, Kokalis SG (1963) J Inorg Nucl Chem 25: 229
307. Schmulbach CD, Derderian C, Zeck C, Sahuri S (1971) Inorg Chem 10: 195
308. Searle HT (1959) Proc Chem Soc London 7
309. Dalgleish WH, Keat R, Porte AL, Tong DA, Ul-Hasan M, Shaw RA (1975) J Chem Soc Daton Trans 309
310. Connolly A, Dalgleish WH, Harkins P, Keat R, Porte AL, Raitt I, Shaw RA (1978) J Magn Reson 30: 439
311. Dalgleish WH, Keat R, Porte AL, Shaw RA (1975) J Magn Reson 20: 351
312. Sridharan KR, Ramakrishna J, Ramachandran K, Krishnamurthy SS (1980) J Mol Struct 69: 105
313. Ahmad KSh, Porte AL (1980) J Mol Struct 58: 459
314. Keat R, Porte AL, Tong DA, Shaw RA (1972) J Chem Soc Dalton Trans 1648
315. Bullen GJ (1971) J Chem Soc A 1450
316. Dougill MW (1963) J Chem Soc 3211
317. Giglio E, Puiliti R (1967) Acta Cryst. 22: 304
318. Faught JB, Moeller T, Paul IC (1970) Inorg Chem 9: 1656
319. Rettig SJ, Trotter J (1973) Can J Chem 51: 1295
320. Galy J, Enjalbert R, Labarre J-F (1980) Acta Cryst 36b: 392
321. Richie RJ, Fuller TJ, Allcock HR (1980) Inorg Chem 19: 3842
322. March WC, Trotter J (1971) J Chem Soc A 169
323. Bandoli G, Casellato U, Gleria M, Grassi A, Montoneri E, Pappalardo GC (1989) J Chem Soc Dalton Trans 757
324. Allcock HR, Allen RW, Bissell EC, Smeltz LA, Telter M (1976) J Am Chem Soc 98: 5120
325. Allcock HR, Stein MT, Bissell EC (1974) J Am Chem Soc 96: 4795
326. Allcock HR, Stein MT (1974) J Am Chem Soc 96: 49
327. Allcock HR, Stein MT, Stanko JA (1970) J Chem Soc Chem Commun 944
328. Oakley RT, Paddock NL, Rettig SJ, Trotter J (1977) Can J Chem 55: 4206
329. Ahmed FR, Singh P, Barnes WH (1969) Acta Cryst 25b: 316
330. Hazekamp R, Migchelsen T, Vos A (1962) Acta Cryst. 15: 539
331. Bullen GJ (1962) J Chem Soc 3193
332. Bovin J-O, Galy J, Labarre J-F, Sournies F (1978) J Mol Struct 49: 421
333. Bovin J-O, Labarre J-F, Galy J (1979) Acta Cryst 35b: 1182
334. Ansell GB, Bullen GJ (1971) J Chem Soc A 2498
335. Begley MJ, Sowerby DB, Tillot RJ (1974) J Chem Soc Dalton Trans 2527
336. Dougill MW (1961) J Chem Soc 5471
337. Pohl S, Krebs B (1975) Chem Ber 108: 2934
338. Bullen GJ (1982) J Cryst Spect Res 12: 11
339. Ritchie RJ, Harris PJ, Allcock HR (1980) Inorg Chem 19: 2483
340. Cordes AW, Swepton PN, Oakley RT, Paddock NL, Ranganathan TN (1981) Can J Chem 59: 2364
341. Allen CW, Faught JB, Moeller T, Paul IC (1969) Inorg Chem 8: 1719
342. Mani NV, Ahmed FR, Barnes WH (1965) Acta Cryst 19: 693
343. Biddlestone M, Bullen GJ, Dann PE, Shaw RA (1974) J Chem Soc Chem Commun 56
344. Pohl S, Krebs B (1976) Chem Ber 109: 2622
345. Babu YS, Manohar H, Shaw RA (1981) J Chem Soc Dalton Trans 599
346. Krishnaiah M, Ramamurthy L, Ramabrahmam P, Manohar H (1981) Z Naturforsch 36b: 765
347. Begley MJ, Sowerby DB, Bamgboye TT (1979) J Chem Soc Dalton Trans 1401
348. Meetsma A., Buwalda PL, Van de Grampel JC (1988) Acta Cryst 44c: 58
349. Enjalbert R, Guerch G, Sournies F, Labarre J-F, Galy J (1983) Z Krist 164: 1
350. Mani NV, Ahmed FR, Barnes WH (1966) Acta Cryst 21: 375

351. Dougill MW, Paddock NL, Sheldrick B (1980) Acta Cryst 36b: 2797
352. Babu YS, Manohar H, Ramachandran K, Krishnamurthy SS (1978) Z Naturforsch 33b: 588
353. Deutsch WF, Gündüz T, Hursthouse MB, Kilic E, Parkes HG, Shaw LS (nee Gözen), Shaw RA, Tüzün M (1986) Phosphorus Sulfur 28: 253
354. Cameron TS, Cordes RE, Jackmann FA (1979) Acta Cryst 35b: 980
355. Bullen GJ, Tucker PA (1972) J Chem Soc Dalton Trans 1651
356. Burr AH, Carlisle CH, Bullen GJ (1974) J Chem Soc Dalton Trans 1659
357. De Ruiter B, Van de Grampel JC, Bolhuis F (1990) J Chem Soc Dalton Trans 2303
358. Bonnett J-P, Labarre J-F (1988) Inorg Chim Acta 149: 187

Biomimetic Chemistry of Hemes Inside Aqueous Micelles

Shyamalava Mazumdar and Samaresh Mitra

Chemical Physics Group, Tata Institute of Fundamental Research,
Homi Bhabha Road, Bombay 400 005, India

Hemes encapsulated in aqueous detergent micelles find themselves in a large macromolecular cavity whose interaction is mainly hydrophobic. It has been suggested that such systems appear to simulate the electrostatic and hydrophobic interactions of the heme cavity in metalloproteins. The present article surveys reported studies on natural and synthetic hemes, both ferric and ferrous, incorporated inside micelles of different sizes and surface charges. The emphasis is laid on multinuclear NMR and optical spectroscopic studies. The effect of micellar interactions on the electronic properties of hemes is discussed and compared with that of the heme cavity in proteins.

1 Introduction 116

2 General Properties of Micelles.. 117

3 Incorporation of Heme Inside Micelles 119

4 Axial Ligation to Heme in Micelles................................... 119
  4.1 Proton Transfer Equilibrium..................................... 119
  4.2 Cyanide Binding 123

5 Electronic Properties of Hemes in Micelles 125
  5.1 Ferric Hemes in Micelles....................................... 125
    5.1.1 High-Spin Hemes 125
    5.1.2 Low-Spin Hemes 129
  5.2 Ferrous Hemes Inside Micelles 132
    5.2.1 Four-Coordinated Iron(II) Hemes 132
    5.2.2 Five-Coordinated Iron(II) Hemes.......................... 136
    5.2.3 Six-Coordinated Iron(II) Hemes 138
  5.3 Linewidth of Heme Proton Resonances Inside Micelles 139
  5.4 Spread of Heme Proton Resonances Inside Micelles.............. 140

6 Structural Disposition of Heme Inside Micelles 141

7 Concluding Remarks.. 143

8 References 144

Structure and Bonding, Vol. 81
© Springer-Verlag Berlin Heidelberg 1993

1 Introduction

Biomimetics of protein pockets in hemoproteins and enzymes is a subject of considerable challenge and continued interest. Electronic properties and bio-chemical functions of the heme prosthetic groups in the hemoproteins have been found to depend on the nature of the protein environment surrounding the heme group [1, 2]. The prosthetic group in cytochromes and many other redox hemoproteins consists of a ferric heme complex, while in oxygen transfer proteins such as myoglobins and hemoglobins the prosthetic group is a ferrous heme [2]. Moreover, the properties of the 'naked' heme in the absence of a protein environment are often quite different from those in proteins [1]. For example, ferrous hemes in aqueous solutions are extremely sensitive to aerial oxidation, whereas myoglobins and hemoglobins consist of stable ferrous heme prosthetic groups [2]. Several studies have suggested that the diversity in the physico-chemical properties of the heme complexes in various proteins largely arises from the hydrophobic interactions and steric effects imparted to the heme moiety by the protein cavity [1, 2].

The model studies on different metalloporphyrin complexes have proposed that the hydrophobicity of the protein pocket has a major influence on the properties of the heme prosthetic group [3–7]. The synthetic and structurally modified natural heme complexes have been extensively studied in order to mimic oxygen binding and many other biochemical properties of hemoproteins. The studies on model heme compounds of 'picket fence', 'basket handle', 'tailed picket fence' porphyrins show that the presence of hydrophobic substituents on the porphyrin ring can stabilize the dioxygen complex of ferrous hemes with axially bound hindered imidazole bases, in both solid as well as non-aqueous solutions [7]. The effects of steric influences on the properties of heme complexes using 'chelated protohemes', 'cyclophane hemes' etc. in solids and solutions have also been investigated [4–6]. These models mimic the hydrophobic and steric interactions of the protein to some extent. However, the studies are usually carried out in the solid state or in non-aqueous media, whereas the biological solvent is water. There is another limitation to the studies on metalloporphyrin as a model in solution. Most metalloporphyrins have a strong tendency to form aggregates in solution [8, 9], which limits their applicability as models. Further, various ferrous heme complexes have been studied in non-aqueous media [10–14]; similar studies in aqueous solutions have, however, been limited because of the possibility of oxidation in aqueous media.

It has been suggested that aqueous micellar systems simulate the electro-static and hydrophobic interactions of the heme cavity [15–23]. Pioneering studies by Simplicio et al. [15–17] have shown that the heme is monodispersed when encapsulated in aqueous micelles. They have studied binding of cyanide and other axial ligands to ferric hemes in micellar environments. These studies [15–23] indicated that a heme encapsulated in an aqueous detergent micelle finds itself inside a large macromolecular cavity whose interactions is primarily hydrophobic, and which is likely to modify its reactivity. They also indicate [15–21] that aqueous micellar solutions can be used to stabilize monomeric model hemes in various oxidation and spin states of the iron with different axial ligands. Recently several detailed and systematic investigations of the biomimetic properties of complexes encapsulated in micelles have been carried out. Micelles of different sizes may mimic heme environments in different proteins. Moreover, the effect of electrostatic interactions as compared to hydrophobic and steric effects can be studied by using heme complexes incorporated in micellar solutions with different surface charges. Finally, the effects of heme paramagnetism on the NMR of the micelles may be used to determine the structural disposition of the heme complexes inside the micellar cavity which can aid understanding of the general properties of the micellar interactions with the paramagnetic macromolecules inside it.

The purpose of this article is to review studies carried out on hemes incorporated inside the micellar cavity, and examine the effect of micellar interaction on the electronic and structural properties of the heme. A comparison of these results with those on the metalloproteins is clearly in order to assess their suitability as models. The article begins with a general introduction to micellar properties, the incorporation of hemes in the micellar cavity, and then discusses results on hemes inside the micelles with different oxidation and spin states, and stereochemistry. The experimental techniques used in the studies on these aqueous detergent micelles are mostly NMR and optical spectroscopy. The present article has therefore a strong emphasis on NMR spectroscopy, since this technique has been used very extensively and purposefully for studies on hemes inside micellar cavities.

2 General Properties of Micelles

Micelles are formed by an aggregation of several surfactant molecules in aqueous solution where the hydrophillic polar head of each surfactant unit is directed towards the aqueous phase in a nearly spherical shape with the non-polar chain groups of the surfactant forming a hydrophobic core. A schematic diagram of the micellar structure is shown in Fig. 1. The phenomenon of micellization by association of surfactant molecules affects the colligative properties e.g. viscosity, surface tension, conductivity, osmotic pressure, NMR chemical shifts and linewidths etc. of the surfactant solutions [24–26]. A sharp change in these properties with increase in the concentration of the surfactants is generally observed during micellization. There are three basic models proposed for micellization, namely *pseudophase, single step equilibrium* and *multistep equilibrium* models. The pseudophase model [24] defines the critical micellar concentration (cmc) as a minimum concentration of the free monomeric surfactant in solution required for the formation of micelles. According to this model the

118

S. Mazumdar and S. Mitra

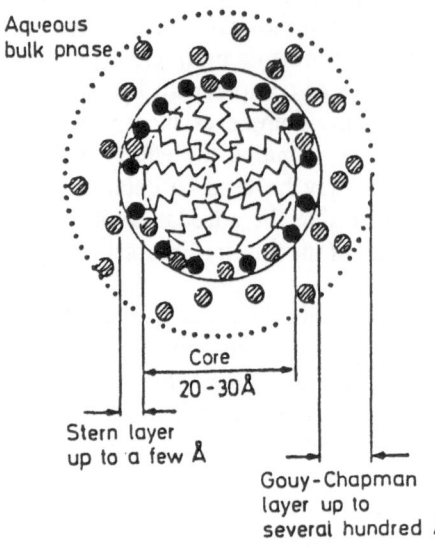

○ WATER MOLECULE

∿● DETERGENT MOLECULE

Aqueous bulk phase

Core
20-30 Å

Stern layer
up to a few Å

Gouy-Chapman
layer up to
several hundred Å

Fig. 1. Schematic diagram of micellar structure

micelles are formed only at surfactant concentrations above the cmc. The single step equilibrium model [24] (also called mass action law model) considers the micellization as a single step equilibrium between the monomer and the micelles formed by aggregation of such monomers. This model has been applied to determine the cmc and aggregation number of many surfactant systems and in most cases the value of cmc obtained by this model is close to that obtained by the pseudophase model.

The multistep equilibrium model [24] considers the coexistence of micelles of different sizes with different aggregation numbers in equilibrium. Thus, this model describes the total micellization process as consisting of several aggregates, and the observed chemical shift of the micellar system is given as $\delta_{obs} = \sum_i iC_i\delta_i$, where δ_i are the chemical shifts of the ith aggregate (or monomer) with concentration C_i. Although the multistep equilibrium model has been shown to provide the most satisfactory explanation of the polydispersity of aggregate sizes, this model becomes difficult to use for micellar solutions containing large aggregates.

Surfactants such as cetyl trimethyl ammonium bromide (CTAB), Triton X-100 (TX-100) and sodium dodecyl sulphate (SDS) are the most commonly used. CTAB forms large micelles [24–26] with aggregation number 61, cmc $\sim 9.2 \times 10^{-4}$ M, and a positive micellar Stern layer; TX-100 has aggregation number 139 with neutral OH groups on the Stern layer, and SDS forms negative micelles with cmc $\sim 8.3 \times 10^{-3}$ M and aggregation number 131. The cmc as well as the average aggregation number can be determined by several experimental methods such as neutron scattering, quasi elastic light scattering, conductivity, osmotic pressure etc. [24–26]. The pseudophase model of micellization has been found convenient for the evaluation of the cmc.

3 Incorporation of Heme Inside Micelles

The details of the preparation of micellar solutions and incorporation of hemes inside the micelles are described in several publications [15–23]. A brief summary is given here. Micellar solutions are generally prepared by dissolving the detergent in aqueous solvent containing 0.1 M tetramethyl ammonium bromide (TMAB), and the resulting suspensions are warmed at 50 °C to give clear solutions [15–17, 27, 28]. Solid samples of the ferric porphyrin complexes are dissolved in micellar solutions and the mixture then allowed to equilibrate at 40–50 °C for ca. 1 hour. In the case of ferric octaethylporphyrin (OEP) complexes, a concentrated acetone solution of the heme was added to the micelles and the resulting solution was heated to remove the actone and to obtain a clear solution of the heme in micelles. Samples prepared in this way have been shown to obey Beer's law over a concentration range of 10^{-8} M to 2×10^{-3} M indicating the presence of monomeric hemin inside the micelles. The pH of the micellar solutions is adjusted with dilute acid or alkali. Solutions at an adjusted pH are again allowed to equilibrate at ca. 45 °C in the dark. To prepare biscyano heme complexes, a concentrated solution of KCN is added to the hemin in micelles at pH ~ 9.6 and equilibrated at 45 °C. The cyano (pyridinato) hemin complex has been prepared by adding a KCN solution to a 18% pyridine-water micellar solution of hemin chloride at a hemin: KCN ratio of 1 : 1.1 (in moles) [29]. For the preparation of ferrous porphyrins inside micelles all operations must be carried out in an inert atmosphere and all reagents (including water) thoroughly deoxygenated. A minimum volume (100 : 1) of a saturated aqueous sodium dithionite has been used to reduce ferric to ferrous. The extent of reduction of ferric to ferrous is estimated by the pyridine hemochrome method [3]. The pyridine hemochrome and the 2-methyl imidazole adducts are also obtained by dithionite reduction of the corresponding ferric complexes encapsulated in the aqueous micelles.

4 Axial Ligation to Heme in Micelles

4.1 Proton Transfer Equilibrium

Acid-base equilibrium in ferric heme complexes inside micelles shows a distinct dependence on the micellar media. The existence of isosbestic points suggests that the following equilibrium process may be operative:

$$[Fe(PP)(L)(H_2O)]^+ \cdot (\text{micelle}) \rightleftharpoons [Fe(PP)(L)(OH)] \cdot (\text{micelle}) + H^+$$

(1)

where 'PP' is the porphyrin dianion and L is an axial ligand. In aqueous solution both metmyoglobin and methemoglobin exhibit equilibrium between the acidic (aqua) and alkali (hydroxo) form. The pKa of this equilibrium is known to be sensitive to the origin of the protein and the nature of the substituents on the heme [2], and ranges from 7.4 to 10.0. The hydrophobic nature of the heme cavity plays an important role in these transitions [2, 20]. Since the micellar cavity is expected to simulate the hydrophobic interaction of the protein cavity, several studies have been carried out to determine the effect of hydrophobic interaction of the micellar cavity on the acid $\rightleftharpoons$ base transition of model hemes [15, 16, 27]. In one such study monomeric bis aqua protoporphyrin ferric complex was stabilised in micelles and pKa of the proton transfer equilibrium was determined. The study gives values of 6.1 in CTAB, 4.7 in TX-100 and 5.5 in SDS, in contrast to a value of 7.1 in the absence of micelles. The studies on the bis aqua complexes of ferric OEP [30] gave a pKa of 4.8. While these studies do show the effect of micellar interaction on the pKa of this transition, the values are not in the range of those observed in the hemoproteins (see Table 1).

   The most elegant example which demonstrates this aspect of proton transfer equilibrium of heme proteins is the study reported on six coordinated aqua (pyridine) iron(III) porphyrin complexes encapsulated in aqueous micelles [27]. The ferric ion in these complexes is axially co-ordinated to a water and a pyridine molecule, thus having a coordination geometry similar to that of the heme in metmyoglobin. The pH dependence of the absorption spectra of the

Table 1. pKa of the aqua $\rightleftharpoons$ hydroxo transition in $[Fe(PP)(py)L]$ (L = H_2O or OH)

Complex	pka(± 0.1)			
	SDS	TX-100	CTAB	Absence of Micelles
[Fe(PP)(py)L]	9.8	9.4	7.7	10.5
[Fe(PPDME)(py)L]	9.2	8.4	7.1	–
[Fe(DP)(py)L]	10.2	10.0	8.6	10.7
[Fe(DPDME)(py)L]	9.7	9.0	8.2	–
[Fe(MsP)(py)L]	10.3	10.1	8.8	10.5
[Fe(MsPDME)(py)L]	9.8	9.1	8.2	–
Fe^{3+} Mb (aplysia)	–	–	–	7.6
Fe^{3+} Mb (Sperm Whale)	–	–	–	9.0

SDS: sodium dodecyl sulphate: CTAB: cetyl trimethyl ammonium bromide; TX-100: Triton X-100; Mb: myoglobin aqua (pyridinato) porphyrin complexes in different micelles and with different heme substituents have been reported [27]. A typical set of pH dependent absorption spectra for the protoporphyrin complex is shown in Fig. 2. At low pH, absorption bands appear around 527 and 556 nm indicating the presence of the aqua form. At higher pH, the spectra show bands around 560 and 600 nm indicating the presence of the hydroxo form. Three isosbestic points are observed, which suggest the presence of only two components, the aqua and the hydroxo forms in equilibrium.

The pKa for the substituted hemins in aqueous pyridine solutions is ~ 10.5 [18] while in the micelles it ranges from 7.1 to 10.3 depending on the micelle and the porphyrin substitution (Table 1). The changes in the pKa between the aqueous pyridine and micellar solutions have been shown not to be associated with any dimer-monomer equilibrium of the heme complexes, since it has been shown that in the concentration range of the reported study the aqua (pyridinato) complexes exit as monomers even in the absence of micelles in the pH range 6.5–12.5. The changes in the pKa have also been shown to be

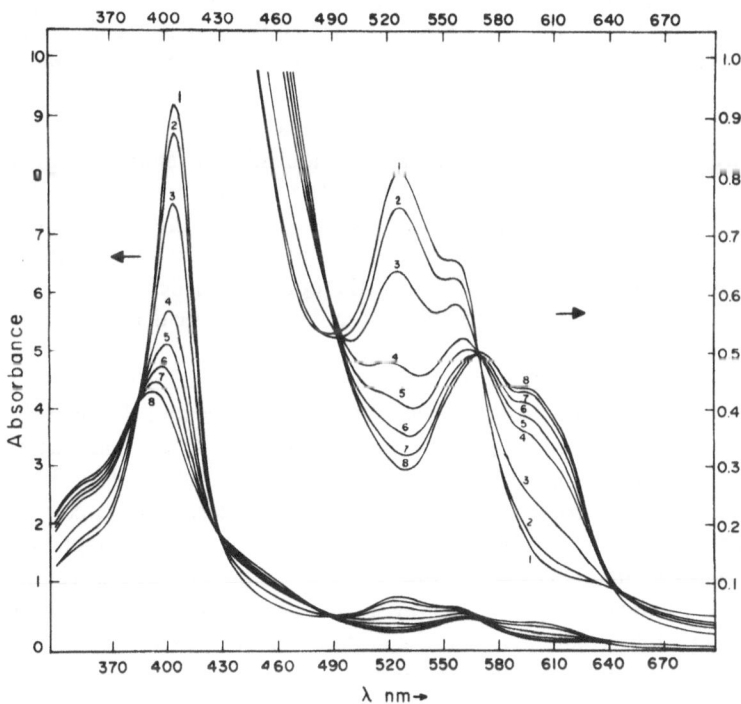

Spectra of Hemin in Pyridine-H_2O in 5% CTAB as a function of pH

Fig. 2. Optical spectra of $[Fe(PP)(py)X]$ $(X = H_2O$ or OH) incorporated in 5% CTAB micelle at different pH; pH = 5.17 (1), 6.82 (2), 7.33 (3), 7.41 (4), 7.79 (5), 8.34 (6), 8.86 (7) and 9.94 (8). (Taken from Ref. 27)

independent of the counterions in the micellar Sterm layer (e.g. NR_3^+ ion in CTAB, SO_4^{2-} in SDS and OH^- group in TX-100).

The changes in pKa between the aqueous solutions and the micelles must therefore arise from the hydrophobic interactions inside the micelle [27, 31]. In general, the pKa in the micelle is lower than that in its absence, implying that the water molecule inside the micelle is more 'acidic' than that outside. The effect of the hydrophobic interaction of the micellar cavity is further evident in the trend of variation of the pKa with micelle (see Table 1). For all the complexes, the pKa increases in the order: anionic SDS > neutral TX-100 > cationic CTAB. The trend is consistent with the expectation based on consideration of the electrostatic charges in the Stern layer. The anionic SDS micelle, in comparison with CTAB and TX-100, stabilizes positive charge on the cationic aqua (pyridinato) ferric heme complexes, and therefore exhibits higher pKa.

The value of the pKa in CTAB is much lower than that in SDS (Table 1). This is true for both the acid and ester forms of the complexes. Further, there is a significant difference in the pKa between the acids and esters of the heme complexes (see Table 1). It is interesting that this difference is largest in the neutral TX-100 (~ 1.0) as compared to that in SDS and in CTAB (~ 0.6). There is also a trend in the variation in pKa with substitution at 2, 4 positions of the porphyrin; i.e. ethyl $\geq$ proton > vinyl (Table 1). This trend is qualitatively the reverse of the trend in the strength of electron withdrawing power of the substituents.

The range of pKa for the acid $\rightleftharpoons$ base equilibrium of aqua (pyridinato) iron(III) porphyrins incorporated in the micelles encompasses the values reported for various ferrihemoglobins and ferrimyoglobins (Table 1). This appears to suggest that the sensitivity of pKa to the structure of the hemoproteins may be simulated by changing the nature of the hydrophobic interactions in the micelles and the substitutions on the porphyrin ring.

The heme peptides obtained from the enzymatic degradation of cytochrome c have been used as a model for hemoproteins because of the structural similarities of the active site [32, 33]. The histidine residue in the peptide chain and the free NH_2 groups of the N-terminal valine and the lysine residues in the undecapeptide (MP11) are potent axial ligands to the heme iron (Fig. 3). The UV-visible spectroscopic studies in aqueous solutions [23, 32–33] show the existence of three proton transfer equilibria in ferric MP11 complex, which are assigned to, (1) binding of histidine to the heme iron by replacing axial water from the diaqua complex (pKa 3.4), (2) replacement of the second water molecule by the α-NH_2 of valine (pKa 5.8) and (3) binding of the ϵ-amino group of lysine by replacement of the valine residue (pKa 7.6). The pKa of the aqua $\rightleftharpoons$ hydroxo transition in MP8 (see Fig. 3) in aqueous solution is 10.0, which is close to the pKa for hemin in aqueous pyridine solution. The nature of the UV-vis spectra of MP11 at ambient pH incorporated inside aqueous micelles is very similar to that of metmyoglobin [2], indicating similarity in their coordination geometry, with fifth and sixth axial sites being occupied by histidine and water respectively. Ferric MP11 in aqueous SDS micellar solution gives a pKa

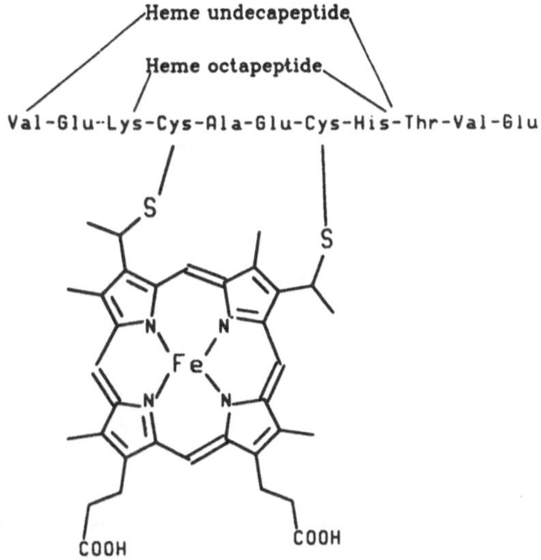

Fig. 3. Schematic diagram of heme peptide

of 7.2 which corresponds to the aqua ⇌ hydroxo equilibrium in the ferric complex [23]. The similarity of this pKa with the acid ⇌ base equilibrium in the [Fe(PP)(py)L], L = H_2O or OH suggests that the MP11 complex in aqueous micelles has a nitrogeneous ligand, possibly the histidine residue axially coordinated to the iron center.

4.2 Cyanide Binding to Heme

Cyanide binding to the hydroxo complex of ferric hemes has been extensively studied in micellar solutions [15–17, 22, 34]. The binding of cyanide is found to be highly dependent on the pH and is maximum at pH ~ 9.6. Optical spectral studies show a distinct isosbestic point at 412 nm for the formation of bis-cyano hemin in SDS from the hydroxo heme. The overall equilibrium is:

$$[(H_2O) Fe(PP)(OH)] \cdot \text{micelle} + 2CN^- \rightleftharpoons [(CN)_2 Fe(PP)]^- \cdot \text{micelle}$$
$$+ OH^-$$

$$K_{eq} = \frac{[[(CN)_2 Fe(PP)]^- \cdot \text{micelle}] [OH^-]}{[[(H_2O) Fe(PP)(OH)] \cdot \text{micelle}] [CN^-]^2}$$

The values of K_{eq} (obtained by spectrophotometric method) are 5.91×10^{-4}, 1.83×10^{-2} and $14.4 \ M^{-1}$ in CTAB, TX-100 and SDS respectively [15, 16]. The decrease in the formation constant of the cyano complex is in the order of increasing hydrophobicity of the micelles. The cyanide complex being ionic is expected to be stabilized in a smaller micelle (e.g. SDS) than in a larger micelle like CTAB. The studies [15, 16, 22] show that in the reaction of cyanide with micelle encapsulated heme, the attacking nucleophile is CN^- (instead of HCN) and the initial uptake of CN^- in the TX-100 and CTAB micelles is very close to those found for metmyoglobins. The formation of isocyano complexes by linkage isomerisation of the biscyano hemin in CTAB and TX-100 has been observed [16]; however this type of isomerisation is not observed in case of SDS micellar solution or in absence of micelles [28]. Spectroscopic studies on binding of cyanate (CNO^-) and imidazole to hemin in micellar solutions have also been reported [17]. The equilibrium constants for the imidazole binding to hemin in micelles are $4.85 \times 10^{-2}\,M^{-1}$, $0.44\,M^{-1}$ and $8.86\,M^{-1}$ for CTAB, TX-100 and SDS respectively. The binding of cyanate in CTAB has an equilibrium constant 1.67×10^{-4}. Imidazole is found to form the bisimidazole species while cyanate forms the mono ligated complex.

The kinetics of binding of cyanide to the heme in micelles has been studied by stopped flow and optical spectroscopy [15, 16]. Two kinetically indistinguishable mechanisms, consistent with the above equilibrium expression, have been proposed for the binding:

Mechanism I

$$Fe(PP)(OH)(H_2O) \cdot micelle + CN^- \underset{k_{-1}}{\overset{k_1}{\rightleftharpoons}} Fe(PP)(CN)(H_2O) \cdot micelle$$

$$+ OH^-$$

$$Fe(PP)(CN)(H_2O) \cdot micelle + CN^- \underset{k_{-2}}{\overset{k_2}{\rightleftharpoons}} Fe(PP)(CN)_2 \cdot micelle + H_2O$$

Mechanism II

$$Fe(PP)(OH)(H_2O) \cdot micelle + CN^- \underset{k_{-1}}{\overset{k_1}{\rightleftharpoons}} Fe(PP)(CN)(OH) \cdot micelle$$

$$+ H_2O$$

$$Fe(PP)(CN)(OH) \cdot micelle + HCN^- \underset{k_{-2}}{\overset{k_2}{\rightleftharpoons}} Fe(PP)(CN)_2 \cdot micelle + H_2O$$

The activation energies calculated for the two steps of the above reaction are $+ 160\,kJ/mol$ for the k_1 step and $+ 78\,kJ/mol$ for the k_2 step [15]. The overall enthalpy of reaction is $- 78\,kJ/mol$. It has been found that the half-life for the k_1 reaction is sensitive to the counterion concentration in case of SDS micelles. The effect of added counterion may be due to the charge neutralisation of the sulphate anion heads in the SDS micellar Stern layer, to facilitate approach and penetration of the CN^- ions at the micelle-water interface. Hemin encapsulated in CTAB micelles reacts much faster with cyanide compared to that in SDS presumably because of the cationic Stern layer in CTAB. The second order rate constant for the first step has been found to be $1.12 \times 10^4 \pm 0.2\,\mathrm{M}^{-1}\mathrm{s}^{-1}$ for CTAB, $5.1 \times 10^2 \pm 0.6\,\mathrm{M}^{-1}\mathrm{s}^{-1}$ for TX-100 and $3.35 \times 10^3 \pm 1.85\,\mathrm{M}^{-1}\mathrm{s}^{-1}$ for SDS micelles at 25 °C [15, 16].

5 Electronic Properties of Heme in Micelles

Both ferric and ferrous hemes have been encapsulated in micelles with different spin states, and coordination geometries and oxidation states have been stabilized. Their electronic and structural properties are reviewed below.

5.1 Ferric Hemes Inside Micelles

5.1.1 High-Spin Hemes

The diaqua and aqua (hydroxo) hemin complexes encapsulated in the micelles [20] are found to be high-spin ($\mu_{\mathrm{eff}} = 5.7 - 5.8\mu_B$). Their high-spin nature is further confirmed from the ESR spectra at 4.2 K (Fig. 4). The spectra are characteristic of high-spin ferric porphyrins with a large zero-field-split $^6A_1$ ground state with $M_S = \pm 1/2$ lying lowest. The spectra are axially symmetric ($g_\parallel = 2.05$, $g_\perp = 6.0$) for the diaqua complex, while for the aqua (hydroxo) complex a rhombic component is observed ($g_1 = 6.1$, $g_2 = 5.6$, $g_3 = 2.0$). The rhombic ESR spectrum in the aqua (hydroxo) complex has been attributed to inequivalent axial ligands which lower the symmetry of the ligand field at the iron site [20]. ESR studies on a high-spin ferric heme complex incorporated in micelles have also been reported [35], which showed a characteristic $g = 6$ resonance at 77 K. Frozen solution Mössbauer spectrum of the aqua complex in SDS micelles shows a quadrupole doublet with values of isomer shift and quadrupole splitting at 78 K being 0.46 mm/s and 1.19 mm/s respectively [36], which are consistent with high-spin character.

The ^{1}H NMR spectra of the diaqua and aqua (hydroxo) hemin complexes encapsulated in micelles have been reported [20] (Fig. 5). The heme methyl resonances in the diaqua species lie in the same region as those of the high-spin bis(dimethyl sulphoxide) iron (III) porphyrin complex [37–39], while those of the aqua (hydroxo) complex appear in a more upfield region. The positions and linewidths of the heme methyl resonances in these complexes are similar to those observed in the aqua and hydroxo hemoproteins [19, 40]. The broadness of the ring methyl resonances of both the diaqua and aqua (hydroxo) species in micelles has been ascribed to arise from the hindered rotational tumbling motion of the heme inside the micelles. The spread and linewidth of these resonances are much larger than those of similar high-spin model heme complexes in simple solution [3].

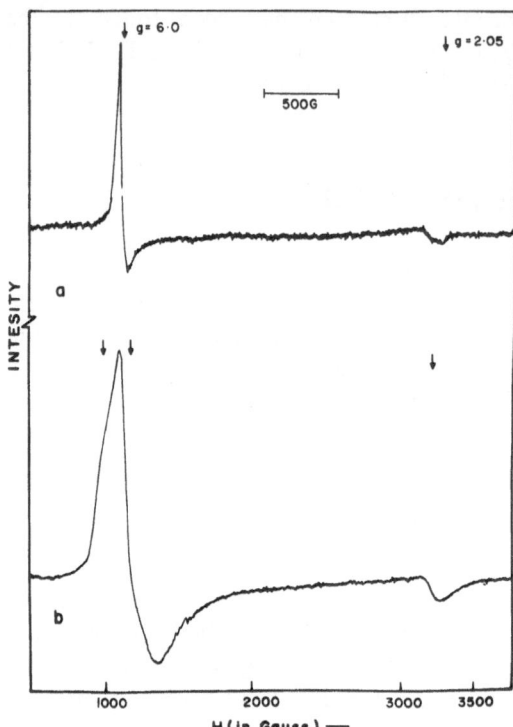

Fig. 4a, b. ESR spectra of frozen solutions at 4.2 K in 5% aqueous sodium dodecyl sulphate micelles of: (a) [Fe(PP)(H$_2$O)$_2$]$^+$ (5 mM);   (b) [Fe(PP)(H$_2$O)(OH)]   (1 mM). (Taken from Ref. 20)

For high-spin ferric porphyrin with axial symmetry, the paramagnetic shift, ($\Delta H/H$) is given by Eq. (2) [3]:

$$\left(\frac{\Delta H}{H}\right) = -\frac{35\bar{g}\beta A}{12(\gamma_N/2\pi)kT} + \frac{28\beta^2}{2k^2T^2}\left(\frac{3\cos^2\theta - 1}{r^3}\right)Df(g) \qquad (2)$$

where f(g) is a function of g tensor and A is the hyperfine coupling constant. D is the zero-field splitting parameter, $(3\cos^2\theta - 1)/r^3$ is the geometric factor and β and γ are the Bohr magneton and nuclear gyromagnetic ratio respectively. $\bar{g}$ is the isotrophic ground state g.

The experimental paramagnetic shift in these compounds is found to obey closely the Curie law of temperature dependence (Fig. 6), suggesting that the contribution of the dipolar term to the paramagnetic shift is negligibly small, and that the paramagnetic shift is predominantly contact in origin [20]. The hyperfine coupling constants of the methyl protons of diaqua and aqua hydroxo complexes have been estimated to be around 0.21 and 0.12 MHz respectively, which are typical of high-spin six-coordinate ferric protoporphyrin complexes [39, 41]. The deprotonation of an axial ligand appears to influence the electronic structure of iron in the porphyrin complexes. Replacement of one of the axial water molecules by a hydroxide ion leads to a paramagnetic shift of 25 ppm

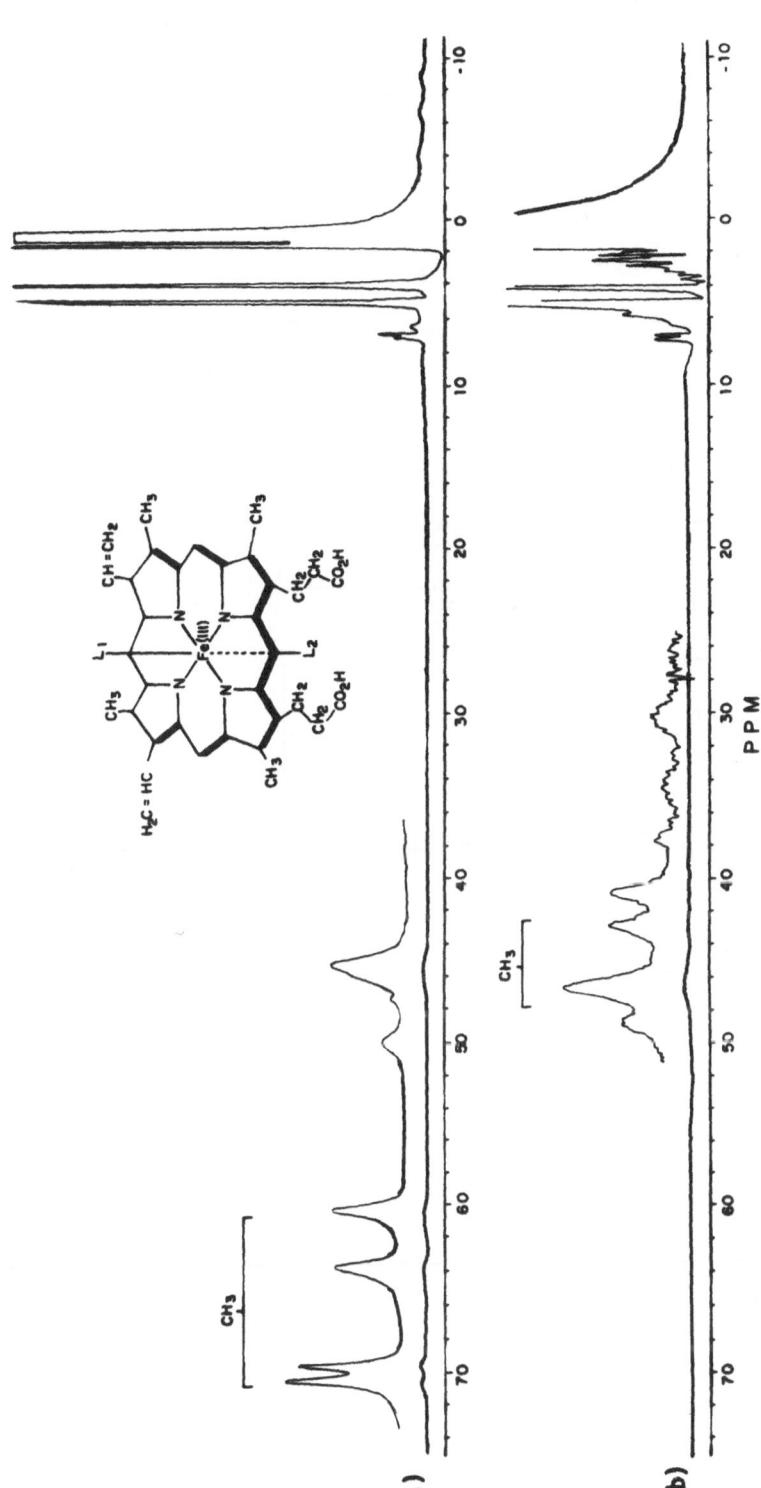

Fig. 5a, b. Proton NMR spectra (500 MHz) at 299 K of (**a**) [Fe(PP)(H$_2$O)$_2$)]$^+$ (1 mM; at pH 2.6) and (**b**) [Fe(PP)(H$_2$O)(OH)] (1 mM; at pH 12.0) in 5% aqueous sodium dodecyl sulphate micelles. Only the heme resonance region has been shown. (Taken from Ref. 20)

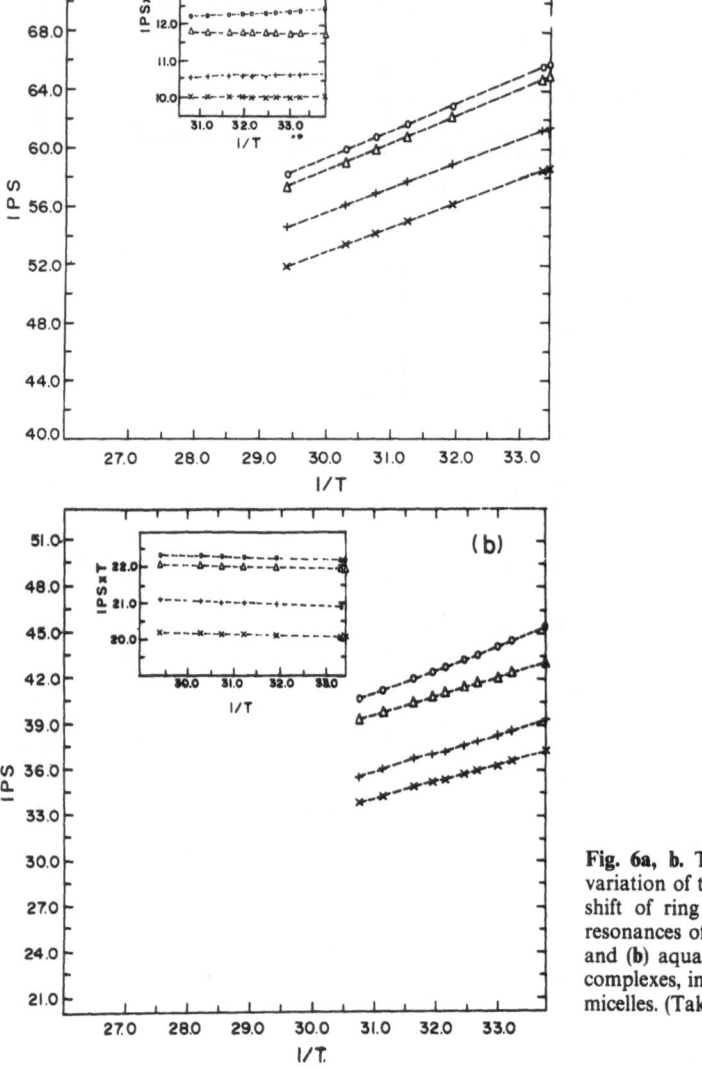

Fig. 6a, b. Temperature variation of the paramagnetic shift of ring methyl proton resonances of (a) diaqua hemin, and (b) aqua hydroxo hemin complexes, in 5% aqueous SDS micelles. (Taken from Ref. 20)

upfield, which is reminiscent of a protein environment [19, 40]. The replacement of H_2O by OH^- in the hemoproteins however produces a change in their ground spin state, while no such change is observed in the micellar systems and other hydroxo ferric porphyrins [19, 40, 42–44]. The proton NMR spectra of the aqua and complexes of ferric octaethyl porphyrin in 5% aqueous SDS solution have also been reported [30]. NMR spectra of these species show close similarity in the chemical shift pattern with other high-spin ferric octaethyl porphyrin complexes in nonaqueous solvents [3, 45, 46]. The meso-H signal in aqua complex appears ~ 10 ppm up field; however in the hydroxo complex it is not resolved possibly due to extensive line broadening. The α–CH_2 in the aqua complex appears at a more down-field region than that in the hydroxo species. The observed paramagnetic shift in the two complexes obeys the Curie law consistent with a high-spin electronic ground state and a large contact contribution to the shift.

The position of the meso-proton NMR signal in ferric octaethylporphyrin complexes has often been considered as diagnostic of the coordination number around the iron center in heme complexes [47]. The five-coordinated high-spin ferric heme complexes [45, 47] show a highly up field (~ -55 ppm) meso-H signal (e.g. [Fe(OEP)Cl]), while in the corresponding six-coordinated species the meso-H signal appears in the down field region (Fe(OEP)(MeOH)$_2$). The appearance of meso-H signal in the up-field region (at -10 ppm) in the aqua OEP complex has been utilized to suggest [30] that the iron atom in this complex resides above the mean porphyrin plane and only one water molecule is axially bound to the iron forming the five-coordinated aqua complex, [Fe(OEP)(H$_2$O)]$^+$.

The pH dependent NMR spectra [23] of the heme peptide complexes inside the SDS micelles show several interesting features (Fig. 7). At very low pH (pH ≤ 1.8) the paramagnetically shifted down field region of the spectra consists of several peaks over a broad envelope in the 55 to 80 ppm range. As the pH is increased above ~ 2, a second set of resonances appears in this region; on further increase in pH the first set of resonances slowly decreases in intensity and finally disappear at pH 5–6. Increase in pH above ~ 7 again causes a rapid decrease in the intensity of the down field signals and a new set of resonances in the 10 to 30 ppm range appears which become quite sharp and prominent at pH ~ 10.

5.1.2 Low-Spin Hemes

The biscyano hemin complexes in different micellar solutions are low-spin and are readily identified by their known visible spectra [15, 16, 17, 28]. The Soret band maxima for biscyano hemin are at 418 nm, 422 nm, 429 nm, 431 nm and those for cyano (pyridinato) hemin are 417 nm, 418 nm, 420 nm, 421 nm in absence of micelles, in SDS, TX-100 and CTAB solutions respectively.

The ^{1}H NMR spectra of the biscyano complex in micelles and in the absence of micelles do not show any significant difference in their general pattern though a small but distinct upfield shift in the heme methyl proton resonances is observed upon incorporation of the heme inside the micelle [22] (Fig. 8). The heme methyl signals for cyano (pyridinato) hemin in aqueous SDS appear further downfield than those for biscyano hemin, which has been attributed to increased asymmetry in the cyano (pyridinato) hemin complex [22]. Further, the upfield shift increases from SDS to CTAB, perhaps due to the change in

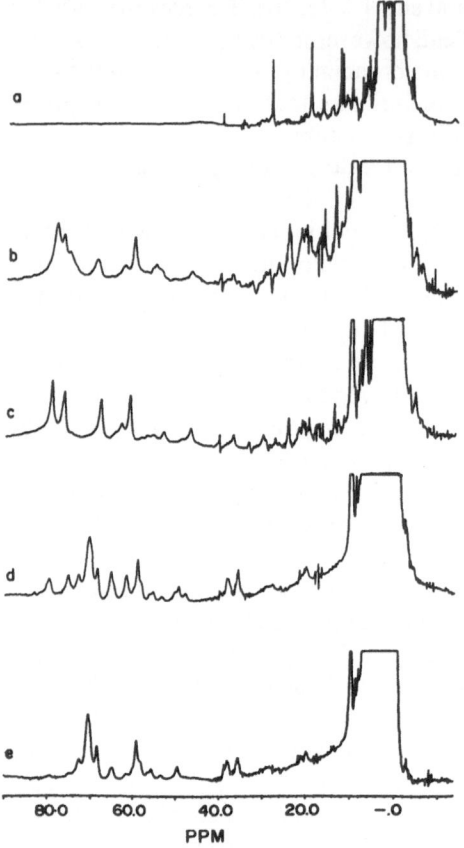

Fig. 7a–e. ^{1}H NMR spectra of heme un-decapeptide (ferric MP11) at different pH. (a) at 9.8 (b) at 6.9 (c) at 5.6 (d) at 2.4 and (e) at 1.8. (Taken from Ref. 23)

hydrogen bonding ability of the 'solvent' inside the micellar cavity compared to that of water. It is known from studies on the biscyano hemins that with decreasing solvent hydrogen bonding ability (i.e. D_2O to DMSO) the heme methyl resonances move upfield from -19.5 ppm to -15.0 ppm [3, 45]. Since CTAB is more hydrophobic than SDS, the change in the paramagnetic shift in the micellar solutions seems to follow the increasing hydrophobicity of the micellar cavity. It is noteworthy that, unlike the high-spin analogues, the NMR spectra of these low-spin complexes do not show any significant increase in the linewidths of the heme methyl signals on incorporation inside the micelles.

The effect of the micelles on the paramagnetic shifts of the heme was very clearly demonstrated [22] in ^{15}N NMR of labelled cyanide in $[Fe(PP)(C^{15}N)_2]^-$ and $[Fe(PP)(py)(C^{15}N)]$ in different micelles as well as in the absence of micelles (Fig. 9). A pronounced systematic downfield ^{15}N shift of the bound cyanide signals is observed on going from a solution without micelles to SDS, to TX-100 and to CTAB micellar solutions which is also the trend in increasing hydrophobicity. The $C^{15}N$ signal is known to be extremely sensitive

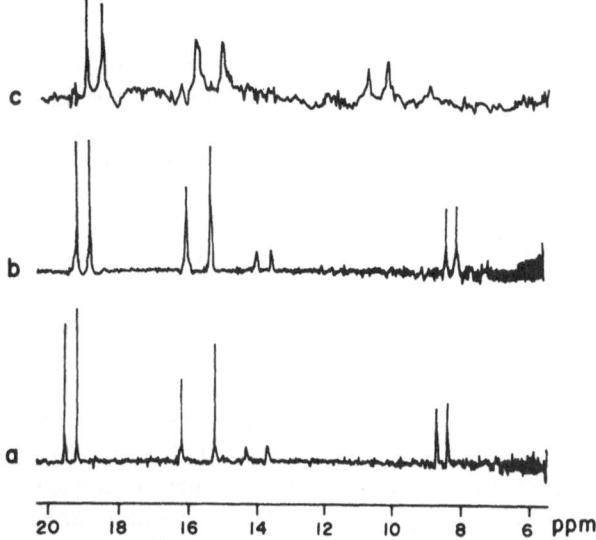

Fig. 8a–c. ^1H NMR spectra of 0.1 mM $[Fe(PP)(CN)_2]^-$ in (**a**) aqueous alkali (pH 9.6); (**b**) 5% SDS micellar solution and (**c**) aqueous 5% CTAB micellar solution. Only the down field heme region is shown. (Taken from Ref. 22)

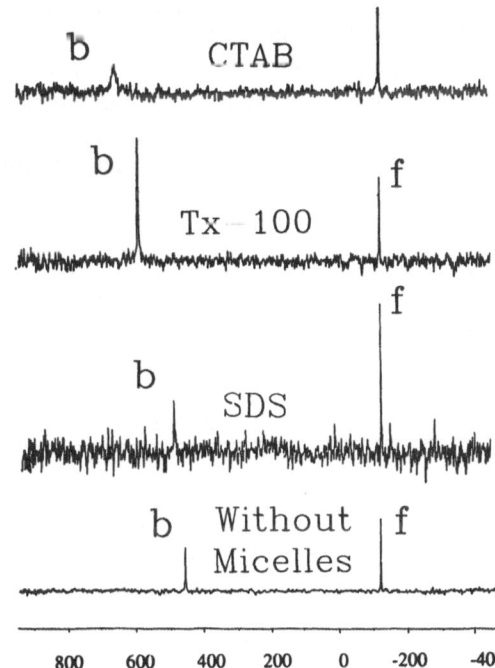

Fig. 9. ^{15}N NMR spectra of 1.2 mM $[Fe(PP)(C^{15}N)_2]^-$ in aqueous alkali (pH 9.6), aqueous SDS, TX-100 and CTAB micellar solutions (Taken from Ref. 22). 'b' denotes N-15 resonance of the bound $C^{15}N$

to the polarity of the solvent [48]. It has been shown that the $C^{15}N$ shift of the bound cyanide in biscyano hemin increases from 448 ppm in D_2O to 732 ppm in DMSO [49]. A similar trend is observed in case of cyano (pyridinato) hemin complexes [22]. Thus the trend in the increase in the down-field shift of bound $C^{15}N$ signals both in biscyano hemin and cyano (pyridinato) hemin in micellar solutions is consistent with the increasing hydrophobic nature of the micelles [17, 34]. The bound $C^{15}N$ signal for cyano (pyridinato) hemin in CTAB appears at 945 ppm, which is close to that for cyano metmyoglobin (931–945 ppm) [50, 51]. The large variation in the $C^{15}N$ paramagnetic shift with micellar size suggests that the spin density at the axial cyanide ligand directly bonded to the iron atom is very sensitive to the micellar environment.

The temperature dependence (296–330 K) of the ring methyl proton resonances of these monomeric heme complexes in the hydrophobic micellar cavity shows [22] a small deviation from the Curie law as in the low-spin complexes in organic and simple aqueous solvents [1, 52]. The origin of such deviation has been variously ascribed [3, 1, 53] either to aggregation or second order Zeeman (SOZ) effect or presence of low-lying spin-quartet state. Since these low-spin hemes in micellar solutions are in deaggregated form, the deviation may be due to the SOZ and/or presence of low-lying excited state.

The NMR spectra of the cyano complex of heme peptides in aqueous SDS micellar solution has also been reported [23]. The nature of the NMR spectra of the cyano ferric MP11 and MP8 complexes is similar though the spread of the heme methyl signals in the MP8 complex is slightly smaller (13.5 ppm) than that in the MP11 complex (16 ppm). This may be due to structural asymmetry arising from the additional three amino acid residues in the peptide chain of the latter complex. Further, the 1- and 3- heme methyl signals in cyano MP11 are closer to each other than in the corresponding MP8 complex, because of the presence of equal number of amino acid residues on both sides of the 2- and 4-porphyrin ring sites in MP11. The heme methyl shifts are reported to obey Curie law of temperature dependence [23].

5.2 Ferrous Hemes Inside Micelles

5.2.1 Four-Coordinate Iron(II) Heme

The dithionite reduction of the micelle encapsulated aqua (hydroxo) ferric hemes at pH ~ 10 (in inert atmosphere) gives an iron (II) porphyrin complex whose optical spectrum [21] shows two well-defined visible bands at 524 and 567 nm and a Soret band split into four bands (Fig. 10). Such spectral features are typical of four-coordinate iron (II) porphyrins. The magnetic moment ($\mu = 3.8 \pm 0.2 \mu_B$) of this sample in the micellar solution is also typical of intermediate spin iron(II) system and is similar to that reported for four-coordinate S = 1 iron(II) porphyrins and phthalocyanine [54–56]. The large orbital contribution ($\mu_{s.o.} = 2.83 \, \mu_B$ for S = 1) observed in this iron(II) porphyrin

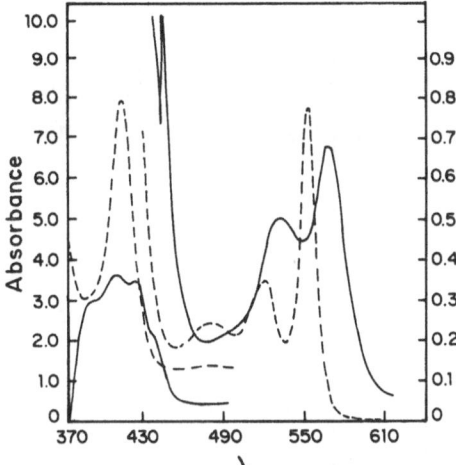

Fig. 10. Electronic spectra of 5×10^{-5} M solution of (a) [Fe^{II}(PP)] (——) and (b) [Fe^{II}(PP)(py)$_2$] (----) in 5% aqueous CTAB micelles at pH 10 (26 °C) (Taken from Ref. 21)

complex is a common feature of all S = 1 four-coordinate iron(II) porphyrins and similar systems [55].

The proton NMR spectra of [Fe(II)(DP)] and [Fe(II)(PP)] in aqueous CTAB micelles (Fig. 11) show general features of the spectra and paramagnetic shift in close agreement with those reported for four-coordinate S = 1 iron(II) porphyrins in benzene [12, 13]. It is important to observe here that such spectra cannot be obtained in simple aqueous solution. Representative chemical shift data for some S = 1 iron(II) porphyrins are included in Table 2. The paramagnetic shift for ring methyl protons obeys closely the Curie law in the temperature range of the study. This is consistent with the S = 1 ground electronic state. Magnetic [54, 55] and NMR studies [14, 57, 58] on several planar iron(II) porphyrins have shown that the ground state of the ferric ion is $^3A_2$ which is extensively mixed with low-lying ^3E and $^3B_1$ states. This mixing causes the commonly observed large orbital contribution to the magnetic moment in the S = 1 iron(II) hemes including those under discussion. The limited magnetic and NMR data in micellar solution are however, inadequate for any detailed probe into the spin-mixing, beyond confirming that these planar iron(II) porphyrins encapsulated in the micelles belong to S = 1 state.

The overall bias of the observed shifts on the in-plane porphyrin resonances in Fig. 11 towards the down field region indicates [12] that the major portion of the paramagnetic proton shift originates from the dipolar interaction. The four-coordinate planar S = 1 iron(II) porphyrins, such as tetraphenylporphyrinato iron(II), [Fe(II)(TPP)] are highly anisotropic [54]. The large orbital contribution to the magnetic moments of these S = 1 iron(II) complexes is consistent with their highly anisotropic character [54]. It is thus expected that the dipolar contribution would be significant and form a major portion of the isotropic shifts for the most protons though its relative magnitude will vary for

Table 2. Chemical shifts (in ppm) for iron (II) hemes

Heme Complex	spin State (S)	Chemical Shifts ppm				
		Ring Me	2–4 H	vinyl CH	meso H	6,7–CH$_2$
[Fe(PP)(2-MeIm)] (in CTAB)	2	22.1, 21.47, 19.4, 6.1	—	12–13	—	13.7
[Fe(PP)(2-MeIm)] (in C$_6$D$_6$)	2	12.8, 12.1, 11.6, 6.9	—	15.5, 11.7 (trans) 10.3, 9.0 (cis)	—	10.7, 10.3
[Fe(DP)(2-MeIm)] (in CTAB)	2	21.53, 20.25, 14.24	46.2, 44.0	—	—	13.5
[Fe(DP)(2-MeIm)] (in C$_6$D$_6$)	2	12.6, 12.3, 12, 7.6	53.55, 49.05	—	—	10.4, 10.1
DeoxyMb^{II} (at pH = 8)	2	16, 11^c	—	—	—	—
DeoxyHb^{II} (at pH = 7)	2	~ 22.5(β), ~ 16, ~ 12	—	—	—	—
[Fe(PP)] (in CTAB)	1	47.9, 47.1, 40, 37.2	—	38	64.5, 62.1	34.3, 33.8
[Fe(PP)] in C$_6$D$_6$	1	46.9, 46.1 40.6, 40.4	—	41.2 39.6	72.6, 69.5 68.4, 64.8	32.2, 31.4
[Fe(DP)] (in CTAB)	1	48.0, 45.05, 43.83, 34.75	—	—	67.8, 66	34, 32
[Fe(DP)] (in C$_6$D$_6$)	1	47.6, 46.1 45.5, 43.9	5.3	—	74.6, 71.7 69.9, 69.6	32.3, 30.9
[Fe(TPP)(2-MeIm)] (in C$_6$D$_6$)	2	—	52.15	—	6.55, 6.59, 6.68^c	—
[Fe(TPP)] (in C$_6$D$_6$)	1	—	12.85	—	20.8, 12.5, 12.5^c	—
[Fe(PP)(py)$_2$] (in CTAB)	0	2.6–3.1^b	—	—	9–10^b	—
[Fe(PPDME)(py)$_2$] (in CDCl$_3$) with N$_2$D$_4$	0	3.5, 3.6	—	—	9.4, 9.56 9.65, 9.7	—

^{a} exact values of the shifts not known; ^{b} Assignments of CH$_3$ and other protons difficult because of masking by resonances from solvent and micelles protons; ^{c} *ortho-*, *meta-* and *para-*phenyl protons respectively

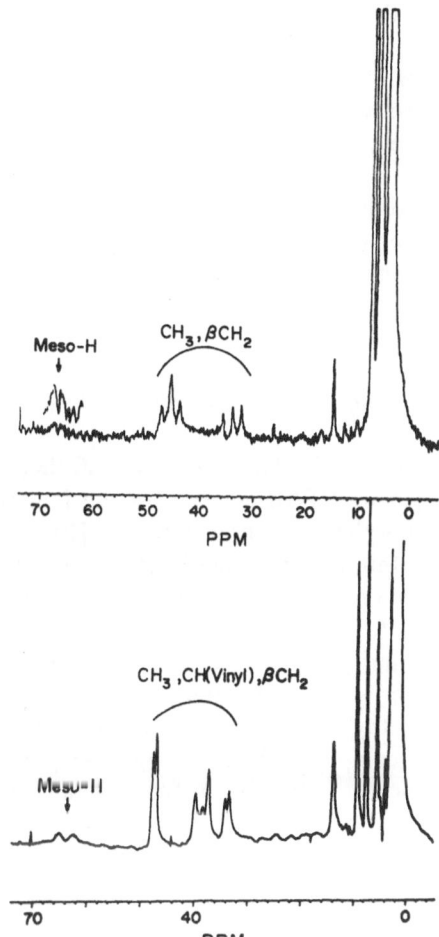

Fig. 11. Proton NMR spectra (500 MHz) at 298 K of the four-coordinate $S = 1$ (*top*) [Fe^{II}(DP)] and (*bottom*) [Fe^{II}(PP)] in 5% CTAB solution (pH 10, Conc. 0.5 mM) (Taken from Ref. 21)

different proton sites. Since the magnetic anisotropy of these systems is known [54, 56] to vary as $1/T$ in the temperature range under discussion, and since the contact term is also expected to vary as $1/T$, the observed Curie behaviour of the paramagnetic shift is consistent with the $S = 1$ spin-state formulation. Frozen solution Mössbauer spectra for the ferrous protoporphyrin complex in CTAB [59] showed isomer shift of 0.57 mm/s and quadrupole splitting of 1.44 mm/s which are characteristic of intermediate spin ($S = 1$) ferrous porphyrin complexes.

The ^{1}H NMR spectra of the Fe(II) porphyrins in aqueous micelles are however characterized by two important differences from those in benzene [12, 13]. The porphyrin proton resonances are much broader in aqueous micelles than in the benzene solution, and the spread of the heme methyl resonances is also larger (see Table 2). These will be discussed in detail in the next sections.

The ferrous complex of octaethyl porphyrin in SDS micelles has been characterized as four coordinated (S = 1) ferrous heme species and is similar to that observed for the ferrous protoheme complex in CTAB. It is noted that ferrous complexes of natural porphyrins cannot be stabilized in aqueous SDS micelles, and much larger aqueous micelles like CTAB were needed to stabilize various ferrous protohemes. This indicates that the environment around the octaethyl porphyrin complex in aqueous SDS is more hydrophobic than that of the analogous natural heme species, suggesting that the OEP moiety is embedded much deeper inside the micellar hydrophobic cavity than the protoporphyrin analogue.

5.2.2 Five-Coordinate Iron(II) Hemes

Addition of a large excess (~ 50 times in moles) of 2-methyl imidazole to an aqueous CTAB solution of the four-coordinate ferrous heme in a nitrogen atmosphere has been reported [21] to give a mono-ligated adduct, [Fe(PP)(2-MeIm)] in the micelle. It can also be obtained by dithionite reduction of an aqueous solution of the ferric 2-methyl imidazole complex in CTAB solution [60, 61]. Both methods yield identical optical spectra (Fig. 12). The close similarity of the spectrum in the visible region to that of deoxymyoglobin (in water) suggests that the complex in the CTAB micelle is a five-coordinate mono 2-methylimidazole complex with sixth coordination site being vacant. Further the complex is high-spin (μ = 4.9 μ_B) like deoxymyoglobin [62]. Addition of 2-methyl imidazole to the aqueous SDS solution of four coordinate ferrous octaethyl porphyrin also gives rise to the mono ligated adduct, [Fe(OEP)(2MeIm)], in the micelle [30]. Its optical spectrum is similar to that of

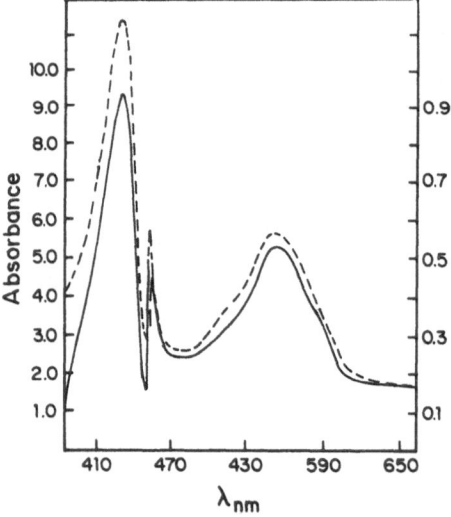

Fig. 12. Electronic spectra of (a) [Fe^{II}(PP)(2-MeIm)] (0.05 mM) in 5% aqueous CTAB micelle (——) and (b) deoxymyoglobin in water (----) at pH 10 (26 °C) (Taken from Ref. 21)

the reported for 2-methyl imidazole adduct of ferrous protoheme complex in aqueous CTAB, suggesting five coordination geometry. The magnetic moment of this complex is $4.8 \pm 0.2 \, \beta$ characteristic of its high-spin (S = 2) nature.

The proton NMR spectra of [Fe(PP)(2-MeIm)] and of [Fe(DP)(2-MeIm)] in aqueous CTAB micellar solution show similarity with that reported for deoxymyoglobin (Fig. 13). Some typical values of the chemical shifts for the high-spin Fe(II) porphyrins in micellar and benzene solutions along with those for deoxymyoglobin are listed in Table 2. The shifts obey Curie law of temperature dependence. The paramagnetic shift is shown to be dominantly contact in origin, and the methyl shift pattern indicates primarily σ-spin transfer mechanism [14]. The dominance of σ spin transfer for the methyl protons is quite reasonable since one of the metal unpaired electrons resides in the $d_{x^2-y^2}$ orbital which is a strongly antibonding type.

Although there is a similarity in the pattern of the isotropic shifts of the high-spin iron(II) porphyrins in the aqueous micellar and benzene solutions, some differences are also noticeable. First, the heme proton resonances in the micelle are much broader than in benzene, and resemble those reported for deoxymyoglobin [62]. Second, the downfield shift of the methyl resonances in

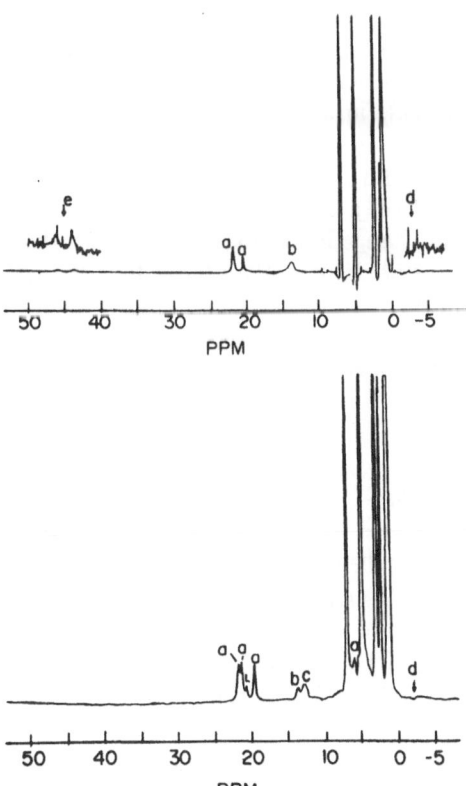

Fig. 13. Proton NMR spectra (500 MHz) at 298 K of the five-coordinate S = 2 (top) [Fe^{II}(DP) (2-MeIm)] and (bottom) [Fe^{II}(PP)(2-MeIm)] in 5% CTAB solution (pH 10, Conc. 0.5 mM) (Taken from Ref. 21)

the micelle (appearing ca. 19 ppm) is much larger than that observed in benzene solution, but is again similar to that observed in deoxymyoglobin. Moreover, as in four-coordinate iron(II) hemes in micelles, the spread of the four methyl protons in these five-coordinate complexes is also found to be larger in micellar solution.

The ferrous MP11 complex in the SDS micellar solution has been stabilized and the absorption spectra resemble closely the five-coordinate high-spin ferrous deoxy myoglobin complex with λ_{max} at $\sim$ 560 and 430 nm [23]. However, in aqueous solution it forms six-coordinate low-spin ferrous heme. It has been suggested that in SDS solutions, Fe(II)-MP11 is coordinated to histidine similar to that in deoxy hemoglobins and deoxy myoglobins [2]. Since the ferrous MP11 is prepared by dithionite reduction of the ferric complex at pH $\sim$ 9.8, the species present at this pH must be mono-histidine mono-hydroxo complex in micellar solution. In the absence of micelles the axial coordination of the ferric heme consists of two nitrogen ligands (histidine and either the valine or the lysine residue) and hence on reduction it gives the six-coordinate low-spin hemochrome species in simple aqueous solution.

The NMR of ferrous complexes of MP11 and MP8 in aqueous SDS solutions has been studied; the spectra are very broad and ill-resolved [23]. The heme proton resonances appear in the range 15 to 30 ppm and resemble those of ferrous hemoproteins. The similarity in the linewidths and spectral range of the heme protons in these ferrous peptide complexes with the ferrous hemoproteins suggests that the larger size of the heme peptide restricts the mobility of the molecule inside the micelles compared to that in case of the protoporphyrin complex in micellar solutions, where the spectrum is better resolved.

Binding of different axial ligands to ferrous hemes have also been studied by NMR [63]. The variation in the nature and position of the heme methyl signals with change in axial ligands suggests that very weak ligand like benzimidazole forms predominantly intermediate-spin (S = 1) complex (μ = 4.0 μ_B) and gradual increase in ligand field strength favours formation of high-spin species. [Fe(PP)(2-MeIm)] with the THF adduct forms a predominantly high-spin ferrous heme complex (μ = 4.4 μ_B), which supports the earlier suggestion [64] of stabilization of high-spin state in six-coordinate bis-THF ferrous hemes.

5.2.3 Six-Coordinate Iron(II) Hemes

The micelle-encapsulated six coordinated bis(pyridinato) iron(II) complexes of protoporphyrin and OEP have been reported by addition of pyridine to the four coordinate ferrous complex in aqueous micellar solution. The optical spectrum of [Fe(II)(PP)(Py)$_2$] in micelle (Fig. 10) is identical to S = 0 six-coordinate bis(pyridinato) iron(II) porphyrin complex [3]. The magnetic moment measurements in solution confirm their diamagnetic nature. The ^{1}H NMR spectra are also characteristic low-spin iron(II) resonances (S = 0) with shifts lying in the diamagnetic region (Table 2).

5.3 Linewidths of Heme Proton Resonances Inside Micelles

A general observation from the ^{1}H NMR studies on micelle-encapsulated heme is that the line width of the heme protons increases significantly in the micellar solutions compared to those in simple solutions. The change in linewidths of heme methyl protons in micelles is not as large in low-spin Fe(III) hemes as in the high-spin ones. The linewidths in four-coordinate and five-coordinate complexes of ferrous hemes in aqueous micellar solutions are also quite broad compared to those observed in benzene solutions [12, 61], but are similar to those of hemoproteins [2, 62]. We examine below the origin and implications of the linewidth change in micellar solutions.

The broadening of the heme proton-resonances is mainly due to the increase in the rotational correlation time of the heme inside micelle. The increase in linewidth, $\Delta\sigma_{obs}$ of a paramagnetic complex inside micelles compared to that in simple aqueous or non-aqueous solution is given as [20–22]:

$$\Delta\sigma_{obs} = [\sigma]_M - [\sigma]_0 \simeq S(S + 1)K_1 (f([\tau_c]_M) - f([\tau_c]_0)) \tag{3}$$

where K_1 is a proportionality constant, σ is the line width, the subscripts M and 0 refer to micellar solution and simple (aqueous or non-aqueous) solutions respectively. Equation 3 shows that the increase in the paramagnetic linewidth of NMR signal of a given proton directly depends on the electron spin (S) of the paramagnetic atom. Moreover, it increases as the total correlation time τ_c of the proton increases. The total correlation time is given by the electron spin relaxation time (τ_s) and the rotational correlation time (τ_r) as, $1/\tau_c = 1/\tau_s + 1/\tau_r$. In the case of high-spin Fe(III) complexes (S = 5/2) the electron spin correlation time τ_s is $\sim 10^{-10}$–10^{-11} s [3] and that for low-spin Fe(III) complexes (S = 1/2) is $\sim 10^{-12}$ s [52]. The typical value of τ_c for paramagnetic iron(II) (i.e., S = 1 or S = 2) porphyrins [3] is $\sim 10^{-13}$ s. The rotational correlation time (τ_r) of the heme in simple solution is ca. 10^{-11} s [52], while τ_r for the micelle protons is ca. 10^{-9}–10^{-10} s [24]. This increase in τ_r in micelles increases the $[\tau_c]_M$ and thus the linewidths of the heme protons in micelles is larger than those in simple non-aqueous solutions. Since the τ_s for high-spin Fe(III) complexes are comparable to the τ_r, the $[\tau_c]_M$ for those systems is increased more significantly by the changes in the value of τ_r leading to a large increase in the linewidth inside micelles. However, the τ_s for the low-spin Fe(III) species is much lower than the τ_r in micelles, thus changes in τ_r due to an increase in viscosity inside the micellar cavity would cause a small increase in the total correlation time (τ_c) for these complexes, and since the total electron spin in low-spin iron is S = 1/2, the increase in linewidths of the heme proton signals in micellar solutions of low-spin heme complexes is much smaller compared to that in the case of high-spin Fe(III) porphyrin (S = 5/2) complexes. Similar reasoning holds good for the significant increase in the paramagnetic linewidths of the heme protons in S = 1 and 2 ferrous heme complexes. Heme peptides in micelles, because of the presence of large peptide side chain, probably have a much larger rotational correlation time than that in case of protoheme complexes. Thus, the heme methyl signals in the heme peptide complexes are relatively large compared to those in the corresponding heme species. As expected, the increase in linewidth in the case of low-spin cyano complexes of the peptides is comparatively smaller than that in the high-spin ferric or the ferrous analogues. The ferrous complexes of heme peptides show extremely broad spectra mainly because of similar reasons. The similarity in the linewidths of the heme methyl signals in the low-spin cyano complexes of ferric hemoproteins e.g. in cyano metmyoglobin and in the high-spin aqua and hydroxo methemoproteins and in high-spin ferrous hemoproteins (S = 2) e.g. deoxymyoglobin (ferrous) [62, 66] with those of the corresponding model heme complexes in aqueous micellar solutions suggest that a similar type of mechanism is responsible for the major contribution to the increase in linewidth in hemoproteins.

5.4 Spread of Heme Proton Resonances Inside Micelles

The second characteristic feature of NMR of the micelle-encapsulated hemes is the spread of the heme methyl resonances. The spread of the heme methyl signals has been a subject of extensive study in both ferric as well as ferrous hemes and hemoproteins [3, 45, 1]. It has been suggested [45, 37] that the spread increases as the symmetry of the porphyrin complex decreases. The range of the heme ring methyl resonances in the high-spin diaqua ferric heme inside aqueous micelles is larger (~ 11 ppm) compared to that in the analogous [Fe(PP)(DMSO)$_2$] in DMSO solution (8.8 ppm) [37]. Similarly, in case of the aqua hydroxo ferric heme in micelles this spread is ~ 9 ppm. In aqua metmyoglobin the spread of the heme methyl resonances is however, much larger (~ 40 ppm) [37] than in any of the hemes in micelles. The spread of the heme methyl signals in cyano metmyoglobin [62] is also more than that in biscyano hemin in aqueous micellar solutions [39, 62, 45] (~ 4 ppm, see Fig. 8) which was found to be slightly less than that in the absence of micelles (see later). The heme methyl signals in high-spin five-coordinate 2-methyl imidazole complex of ferrous hemes in micelles show a larger spread (~ 10 ppm) compared to that in benzene solution (~ 6 ppm) (Table 2). Similarly, in case of four-coordinate intermediate-spin ferrous heme inside micelles, the spread of these methyl resonances is larger compared to that in benzene solution. The spread of heme methyl signals in cyano (pyridinato) hemin in aqueous micelles (~ 7.5 ppm) is more than that in simple pyridine-water solution (~ 6 ppm) with a small downfield shift of the two low-field methyl signals and a small upfield shift of the other two methyl peaks (Fig. 8).

These observations suggest that the environment of heme inside the micellar cavity is more asymmetric than that in simple solution; this asymmetry is much larger inside the protein cavity [45]. An unsymmetrical structural disposition of the heme complex inside the micellar cavity has been proposed [67]. Besides, the increase in the spread of the heme methyl signals in the five-coordinate 2-methyl imidazole ferrous heme complex might also arise from the decrease in symmetry due to restricted orientation of the axial ligand [45], though this contribution is expected to be small because of the possible free rotations of the 2-methyl imidazole moiety about the Fe–N bond in the temperature range of such studies. Thus the increase in the spread of the heme methyl signals in high-spin Fe(II) hemes may mainly be due to the porphyrin-micelles interactions. The peptide side chain in heme peptide complexes causes a significant increase in the in-plane asymmetry of these complexes. Hence, while the spread of the heme methyl shifts in low-spin cyano peptide complexes ranges between 14–16 ppm this spread is only 4–8 ppm in the cyano ferric heme complexes.

There is also a possibility of changes in the in-plane asymmetry of the porphyrin ring due to aggregation in simple solutions [9]. This may introduce a change in the spread of the heme methyl signals. However, since in high-spin ferric heme complexes the extent of aggregation is small and the aggregates formed are symmetric in nature, the heme methyl signals in high-spin hemins may not have any extra spread due to aggregation. The low-spin cyano complexes of ferric hemes are known to form asymmetric slip-over type aggregates [68, 69], which may cause an extra spread in the heme methyl signals in the low-spin biscyano hemin in aqueous alkali solution. Because of this effect of aggregation the spread of the heme methyl signals in low-spin hemins in simple aqueous solution is slightly larger than that in micellar solution, and similar trend is also observed in the case of cyano complexes of heme peptides [23].

6 Structural Disposition of Heme Inside Micelles

The presence of paramagnetic iron porphyrin complexes inside the micelles shows distinct effects on the NMR spectra of the micellar protons [67] (Fig. 14). The micellar proton signals show a small downfield shift in the presence of 0.1 mM biscyano hemin in SDS while the aqua (hydroxo) hemin (0.1 mM) causes an upfield bias. These changes in the chemical shifts of the micellar protons due to the presence of paramagnetic molecule have been attributed to dipolar interactions between the micellar protons and the iron center [3]. However, because of the broadness and poor resolution of the SDS proton resonances (Fig. 14) proton relaxation time measurement is not suitable for deducing structural information in micelles [67]. ^{13}C NMR spin-lattice relaxation time measurement of the surfactant molecules was therefore used to determine the structure of the micelles in solution [67–71]. The presence of paramagnetic ferric heme complexes affects the ^{13}C chemical shift of SDS micelles to a very small extent. However, the spin-lattice relaxation rates for SDS carbons show a significant systematic variation with increasing concentration of the hemin complex. The paramagnetic spin-lattice relaxation time T_{1M} was experimentally determined for different carbon atoms of the detergent molecule. T_{1M} is related [67] to the average distance of the relaxing nucleus (carbon) from the paramagnetic

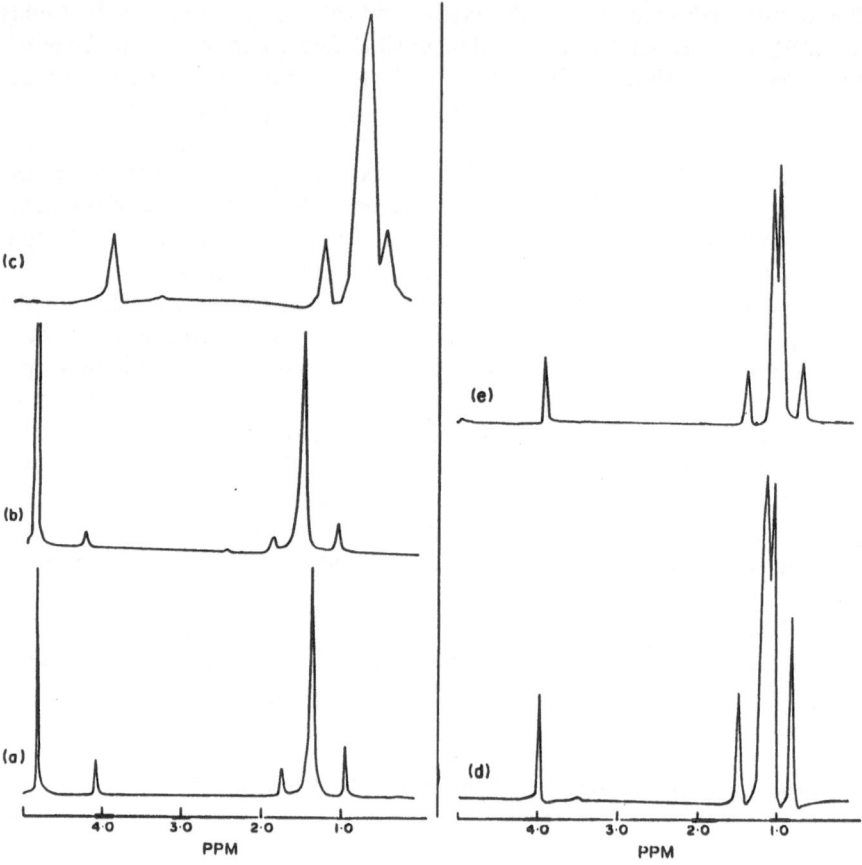

Fig. 14a–e. Proton NMR spectra of 5% aqueous sodium dodecyl sulphate (a) in aqueous solution in absence of any paramagnetic complex, (b) in presence of biscyano hemin (0.1 mM), (c) in presence of aqua (hydroxo) hemin, (d) in presence of hydroxo (pyridinato) hemin and (e) in presence of cyano (pyridinato) hemin (Taken from Ref. 67)

ion by Solomon-Bloembargen equation [24, 72] given as:

$$\frac{1}{T_{1M}} = \frac{2}{15} \frac{\gamma_I^2 g^2 S(S+1)\beta^2}{R_{IS}^6} \left(\frac{3\tau_C}{1 + \omega_I^2 \tau_C^2} + \frac{7\tau_C}{1 + \omega_S^2 \tau_C^2} \right)$$
$$+ \frac{2}{3} S(S+1) \left(\frac{A}{\hbar} \right)^2 \left(\frac{\tau_e}{1 + \omega_S^2 \tau_e^2} \right) \tag{4}$$

Assuming a smooth spherical structure of the micellar surface [73] with a uniform density distribution, the average distances of the micellar carbon and the ironcenter of the porphyrin complex were evaluated [67] and the disposition of porphyrin molecule in the micelle was determined (Fig. 15). The propionic acid side chains in the heme molecule have been proposed [67] to be directed

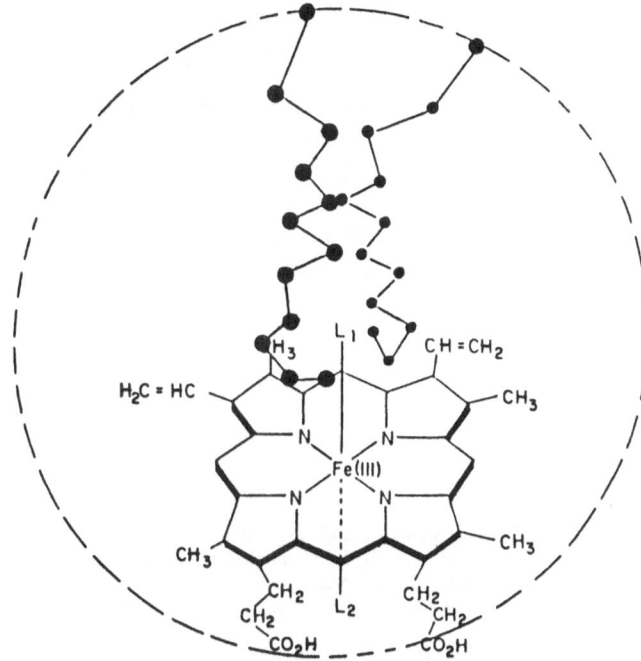

Fig. 15. Schematic diagram of the disposition of ferric porphyrin complex inside micelle. (Taken from Ref. 67)

towards the micellar Stern layer, and the rest of the porphyrin ring resides inside the micelles. The results also suggest a radial distribution of the terminal carbons near the core of the micelles. A very similar structure of micelles was also deduced by Cabane [70] by using the relaxation probe outside the micellar surface. Similar studies with the hydroxo complex of ferric octaethylporphyrin [30] also show a very similar distribution of detergent atoms inside the micellar cavity. However, because of high symmetry and hydrophobic nature, the octa-ethylporphyrin ring is embedded near the center of the micellar cavity.

7 Concluding Remarks

The results presented in this article on the ferric and ferrous hemes in different aqueous detergent micelles have brought out several interesting features. In general, it can be observed that ironporphyrin complexes exist in monomeric form inside the micellar cavity. This is a unique advantage since it provides an opportunity to study these hemes in monomeric form in an aqueous medium without any of the usual complications of aggregation. The ferric and ferrous hemes in different stereochemistries and spin-states are stable in the micellar cavity. Of particular interest is the somewhat higher stability of iron(II) hemes in the micelles. The NMR studies of hemes inside the micelles appear to provide valuable information on the origin of hemeprotein NMR resonances. There is a resemblance in the nature of their chemical shift, linewidth and spread of the proton resonances which, though of limited scope, is instructive. The dominant effect of the hydrophobic interaction of the micellar cavity is evident in most studies. It would be useful to carry out photochemical and redox reaction studies on hemes inside micelles as an aid to understand similar studies on hemeproteins.

References

1. Wüthrich K (1973) Structure and Bonding 8: 53
2. Antonini E, Brunori M (1971) In: Hemoglobins and myoglobins in their reactions with ligands, North-Holland, Amsterdam, chap 4
3. LaMar GN, Walker FA (1979) In: Dolphin D (ed) The porphyrins. Academic, New York, vol IV, p 61
4. Chang CK, Traylor TG (1973) J Am Chem Soc 95: 5810
5. Traylor TG, Chang CK, Geibel J, Berzinis A, Mincey T, Cannon J (1979) J Am Chem Soc 101: 6716
6. Traylor TG, Campbell D, Tsuchiya S, Mitchell M, Stynes DV (1980) J Am Chem Soc 102: 5959
7. Collman JP, Reed CA (1973) J Am Chem Soc 93: 2048
8. White WI (1978) in: Dolphin D. (ed.) 'The Porphyrins', Academic, New York, 5, 303
9. Mazumdar S, Mitra S (1990) J Phys Chem 94: 561
10. (a) Brault D, Rougee M (1974) Biochemistry 13: 4591; (b) Brault D, Rougee M (1975) Biochemistry, 14: 4100
11. Brault D, Rougee M (1973) Nature (London), 241: 19
12. Goff HM, LaMar GN, Reed CA (1977) J Am Chem Soc 99: 3641
13. Mispelter J, Momenteau M, Lhoste JM (1977) Mol Phys 33: 1715
14. (a) Goff HM, LaMar GN (1977) J Am Chem Soc 99: 6599; (b) LaMar GN, Budd DL, Goff HM (1977) Biochem Biophys Res Comm 77: 104
15. Simplicio J (1972) Biochemistry 11: 2525; Simplicio J (1972) Biochemistry 11: 2529
16. Simplicio J, Schwenzer K (1973) Biochemistry 12: 1923
17. Simplicio J, Schwenzer K, Maenpa F (1975) J Am Chem Soc 97: 7319
18. Bartocci C, Sc,ola F, Ferri A, Carassiti V (1979) Inorg Chim Acta 37: L473
19. McGratth TM, LaMar GN (1978) Biochim Biophys Acta 534: 99
20. Mazumdar S, Medhi OK, Mitra S (1988) Inorg Chem 27: 2541
21. Medhi OK, Mazumdar S, Mitra S (1989) Inorg Chem 28: 3243
22. Mazumdar S, Medhi OK, Mitra S (1990) J Chem Soc (Dalton) 1057
23. Mazumdar S, Medhi OK, Mitra S (1991) Inorg Chem 30: 700
24. Chachaty C (1987) Prog NMR Spectroscopy 19: 183
25. Wennerström H, Lindman B (1979) Physics Reports 52: 1
26. Lindman B, Wennerström H (1980) In: Topics in Current Chemistry 1
27. Mazumdar S, Medhi OK, Kannadagulli N, Mitra S (1989) J Chem Soc (Dalton) 1003
28. Hambright P, Chock PB (1975) J Inorg Nucl Chem 37: 2363
29. Caughey WS, Barlow CH, O'Keefe DH, O'Toole MC (1973) Ann NY Acad Sci 206: 296
30. Mazumdar S (1991) J Chem Soc (Dalton) 2091
31. Bunton CA, Romsted LS, Sepulveda L (1980) J Phys Chem 84: 2611
32. (a) Tsou CL (1951) Biochem J 49: 362; (b) Tuppy H, Palèus S (1955) Acta Chem Sc; 9: 353; (c) Palèus S, Ehrenberg A, Tuppy H (1955) Acta Chem Scand 9: 365

33. Jehanli AMT, Stotter DA and Wilson MT (1976) Eur J Biochem 71: 613
34. Minch MJ, LaMar GN (1982) J Phys Chem 86: 1400
35. Smith TD, Gaunt R, Ruzic I (1983) Inorg Chim Acta 78: 103
36. Medhi OK, Houlton A, Silver J (1989) Inorg Chim Acta 161: 213
37. Budd DL, LaMar GN, Langry KC, Smith KM, Nayyir Mazhir R (1979) J Am Chem Soc 102: 6091
38. Zobrist M, LaMar GN (1978) J Am Chem Soc 100: 1944
39. Kurland RJ, Little RG, Davis DG, Ho C (1971) Biochemistry 10: 2237
40. Morishima I, Ogawa S, Inubushi T, Yonezawa T, Iizuka TT (1977) Biochemistry 16: 5109
41. Behere DV, Birdy R, Mitra S (1984) Inorg Chem 23: 1978
42. Chang L, Latos-Grzynski, Balch AL (1982) Inorg Chem 21: 2412
43. Fielding L, Eaton GR, Eaton SS (1985) Inorg Chem 24: 2309
44. Shirazi WA, Bruice TC (1986) Inorg Chem 25: 3845; and references therein
45. Goff HM (1983) In: Lever ABP, Gray HB (eds) Iron porphyrins. Addison-Wesley, Reading MA, part I chap 4, p 239
46. Morishima I, Katagawa S, Matsuki E, Inubushi T (1980) J Am Chem Soc 102: 2429
47. Behere DV, Goff HM (1984) J Am Chem Soc 106: 4946
48. Morishima I, Inubushi T (1977) J Chem Soc Chem Comm 616
49. Morishima I, Inubushi T (1968) J Am Chem Soc 100: 3568
50. Morishima I, Inubushi T, Neya S (1977) Biochem Biophys Res Comm 78: 739
51. Morishima I, Inubushi T (1978) Biochem Biophys Res Comm 80: 199
52. von Goldammer E, Zorn H, Daniels A (1975) Eur J Biochem 57: 291
53. Horrocks Jr DW, Greenberg RG (1974) Mol Phys 27: 993
54. Boyd PWD, Buckingham DA, Meeking RF, Mitra S (1979) Inorg Chem 18: 3585
55. Mitra S (1983) In Lever ABP, Gray HB (eds) Iron Porphyrins. Addison-Wisley, Massachusetts, Part II, page 1
56. Barraclough CG, Martin RL, Mitra S, Sherwood RC (1970) J Chem Phys 53: 1643
57. Mispelter M, Momenteau M, Lhoste JM (1981) Biochimie 63: 911
58. Mispelter M, Momenteau M, Lhoste JM (1980) J Chem Phys 72: 1003
59. Medhi OK, Silver J (1989) J Chem Soc (Chem comm) 1199
60. Ann Walker F, Lo MW, Ree MT (1976) J Am Chem Soc 98: 5552
61. Satterlee JD, LaMar GN, Frye JS (1976) J Am Chem Soc 98: 7275
62. LaMar GN (1979) In: Shulman RG (ed) Biological Applications of Magnetic Resonance. Academic, New York, chap 7, p 305
63. Mazumdar S, Medhi OK (1990) J Chem Soc (Dalton) 2633
64. (a) Sams JR, Tsin TB (1974) Chem Phys Lett 25: 599; (b) Collman JP, Reed CA (1973) J Am Chem Soc 95: 2048; (c) Dolphin D, Sams JR, Tsin TB, Wong KL (1976) J Am Chem Soc 98: 6970
65. Behere DV, Birdy R, Mitra S (1982) Inorg Chem 21: 386
66. Sheard B, Yamane T, Shulman RG (1970) J Mol Biol 53: 35
67. Mazumdar S (1990) J Phys Chem 94: 5947
68. Viscio DB, LaMar GN (1978) J Am Chem Soc 100: 8092
69. Viscio DB, LaMar GN (1978) J Am Chem Soc 100: 8096
70. (a) Cabane B, Physique J (1981) 42: 847; (b) Chevalier Y, Chachaty C (1985) J Phys Chem 89: 875
71. (a) Bert-Ove Persson, Drakenberg T, Lindman B (1976) J Phys Chem 80: 2174; (b) Bert-Ove Persson, Drakenberg T, Lindman B (1979) J Phys Chem 83: 3011; (c) Williams E, Sears B, Allerhand A and Cordes EH (1973) J Am Chem Soc 95: 4871
72. (a) Dwek RA (1973) In: Nuclear magnetic resonance (NMR) in biochemistry, Clarendon Oxford; (b) Dwek RA (1974) In: Sigel H (ed) Metal ions in biological systems, vol 4; Marcel Dekker New York
73. Gruen DWR (1985) J Phys Chem 89: 153

Chemistry Relevant to the Biological Effects of Nitric Oxide and Metallonitrosyls

M. J. Clarke and J. B. Gaul

Department of Chemistry, Boston College, Chestnut Hill, Massachusetts 02167, USA

Nitric oxide has recently been found to be important messenger molecule mediating critical physiological functions such as neurotransmission and muscle relaxation. The nitroprusside ion, which is clinically used as a vasodilator, functions by releasing NO in vivo. The reactivities of these molecules are surveyed with a view toward the development of metallonitrosyls as pharmaceutical agents.

1 Introduction 148

2 Chemistry of Nitric Oxide.. 148

3 Medical Use of Sodium Nitroprusside 151

4 Biological Functions of Nitric Oxide 152
    4.1 Smooth Muscle Relaxation and Vasodilator Functions of Nitric Oxide 152
    4.2 Messenger Role of NO in the Central Nervous System 154
    4.3 Nitric Oxide as a Macrophage Defense Weapon.................................... 156
    4.4 Nitric Oxide Toxicity.. 156

5 Reaction Chemistry of Nitroprusside and Related Metallonitrosyls 157
    5.1 Overview of Reactions of Nitroprusside 157
    5.2 Redox Reactions of Nitroprusside....................................... 160
    5.3 Correlations of Metallonitrosyl Properties with the Lever Parameter 161
    5.4 Reactions of Nitroprusside with Hemoproteins 164
    5.5 Photochemistry of Nitroprusside 165
    5.6 Substitution Reactions on Nitroprusside 166
    5.7 Reactions of Coordinated Nitric Oxide 167
    5.8 Nitroprusside Oxidation of Protein Sulfhydryls......................... 170

6 Enzymatic Interactions of Nitric Oxide 170
    6.1 Reactions of NO with Hemes and Guanylate Cyclase............................ 170
    6.2 Nitric Oxide Synthase... 173
    6.3 Nitric Oxide Production by Denitrifying Bacteria 175

7 Bacteriostatic Effects 176

8 Conclusion... 176

9 References 177

Note Added in Proof.. 181

1 Introduction

The advent of metallonitrosyl drugs patterned after the well known transition metal complex, nitroprusside [1], is reminiscent of the emergence of the platinum antitumor drugs, which derive from a serendipitous discovery concerning another historic inorganic complex [2]. Sodium nitroprusside was first characterized almost 150 years ago [3, 4]. Its effectiveness as a blood-pressure lowering agent, which has been known since 1929 [5], arises from its ability to dilate arteries. The clinical use of nitroprusside as a potent hypotensive agent now spans three decades [6]; however, only recently has its marked pharmacological effect been attributed to its donation or release of nitric oxide (NO) [7]. This smallest, thermally stable, odd-electron molecule [8], also occurs in biology as an intermediate in the bacterial reduction of nitrite [9] and in the body, where it is generated by the enzyme, nitric oxide synthase [10]. Since its biological half-life is brief (~ 5 s) [11, 12], it is somewhat surprising that this toxic and often polluting, inorganic molecule serves several important and diverse biological functions [13]. Iron nitrosyl complexes and nitric oxide have also been implicated in the action of nitrite as a food preservative [14]. This review surveys the literature relevant to the biological effects of NO and metallonitrosyl complexes through July 1992 from the coordination chemist's perspective.

2 Chemistry of Nitric Oxide

Nitric oxide has the electronic configuration $\sigma^2\sigma^{*2}$, $\sigma^2\pi^4\pi^*$ with a net bond order of 2.5, a net bond strength of 627.5 kJ/mol, an interatomic distance of 1.15 Å, and an infrared stretching frequency of 1840 cm^{-1}. It is thermodynamically unstable ($\Delta G^\circ = 86.57$ kJ/mol, $\Delta S^\circ = 217.32$ kJ/mol K) and decom-

Table 1. Reduction potentials of NO, NO$_2$ and related compounds [157, 158]

Reaction in 1 M Acid	E (V)	Reaction 1 M Base	E (V)
$2NO + 4H^+ + 4e \rightarrow N_2 + 2H_2O$	1.678	$2NO + 2H_2O + 4e \rightarrow N_2 + 4OH^-$	0.850
$2NO + 2H^+ + 2e \rightarrow N_2O + H_2O$	1.587	$2NO + H_2O + 2e \rightarrow N_2O + 2OH^-$	0.759
$2NO + 2H^+ + 2e \rightarrow H_2N_2O_2$	0.71	$2NO + 2e \rightarrow N_2O_2^{2-}$	0.18
$NO + H^+ + e \rightarrow NOH$	-0.54	$NO + e \rightarrow NO^-$	-0.81
$NO_2 + H^+ + e \rightarrow HNO_2$	1.108	$NO_2 + e \rightarrow NO_2^-$	0.917
$HNO_2 + H^+ + e \rightarrow NO + H_2O$	0.984	$NO_2^- + H_2O + e \rightarrow NO + 2OH^-$	0.481
$NO_3^- + 4H^+ + 3e \rightarrow NO + 2H_2O$	0.955	$NO_3^- + 2H_2O + 3e \rightarrow NO + 4OH^-$	0.149
$NO_3^- + 3H^+ + 2e \rightarrow HNO_2 + H_2O$	0.940	$NO_3^- + H_2O + 2e \rightarrow NO_2^- + 2OH^-$	0.017
$NO_3^- + 2H^+ + e \rightarrow NO_2 + H_2O$	0.773	$NO_3^- + H_2O + e \rightarrow NO_2 + 2OH^-$	-0.883

poses to N_2 and O_2 at high temperature [15, 16]. The unpaired π^* electron ($\mu = 1.837\,\beta$), which is fairly evenly distributed over the two atoms, is relatively easily lost ($\Delta H_{ion} = 890.6$ kJ/mol) to give $N = O^+$ (bond length 1.06 Å, $\nu NO = 2150$–2400 cm^{-1} [15]. Its low dipole moment (0.15D) and Ostwald partition coefficients (water, 0.074 at $P_{NO} = 1$; toluene, 0.318; methanol, 0.363; pentane, 0.26 at 0 °C) indicate that it is relatively lipophilic [17].

Nitric oxide is capable of undergoing a variety of redox reactions, which are summarized as reduction potentials in Table 1, including disproportionation to N_2O and NO_2 (N_2O_3) or even to N_2 and O_2 with a zeolite catalyst [8]. At low temperature it can dimerize in a side-to-side fashion to form N_2O_2 (C_{2v}) and, at higher temperature, pair electrons with NO_2 to form N_2O_3 [18].

$$3NO \rightarrow N_2O + NO_2$$

$$\Delta G° = -104.1\,\text{kJ/mol}, \qquad \Delta H° = -195.2\,\text{kJ/mol}$$

$$4NO \rightarrow N_2O + N_2O_3$$

$$\Delta G° = -102.8\,\text{kJ/mol}, \qquad \Delta H° = -155.4\,\text{kJ/mol}$$

$$NO_2(g) + NO(g) \rightarrow N_2O_3(g)$$

$$\Delta G° = 1.59\,\text{kJ/mol}, \qquad \Delta H° = -40.5\,\text{kJ/mol},$$

$$\Delta S° = 139\,\text{J/mol K}, \qquad K = 1.91\,\text{atm} \; [8]$$

$$NO_2(aq) + NO(aq) \rightleftarrows N_2O_3(aq)$$

$$k = 1.1 \times 10^9, K - 5 \times 10^4 \; [18]$$

$$N_2O_3(aq) + H_2O \rightarrow 2HNO_2(aq)$$

N_2O_3 is a planar molecule with a long ON–NO_2 bond (1.864 Å) between the sp^2 nitrogens [19]. It hydrates in water to form HONO in acidic solution or nitrite in basic media. Reaction with H_2SO_4 or other concentrated acids afford a route to nitrosonium salts, such as $NO[HSO_4]$ [8].

The electrochemical oxidation of NO around 1.0 V (cf. Table 1) has recently been used to devise NO-selective amperometric microprobe electrodes to detect its release in biological tissues [20, 21]. In aqueous solution, NO is quickly air-oxidized to NO_2. Surprisingly, the kinetics of this reaction have not been investigated. In the gas phase at room temperature, NO is rapidly oxidized by oxygen ($\Delta G° = -70.54$ kJ/mol) in a third-order mechanism [8]:

$$2NO(g) \rightleftarrows N_2O_2(g)$$

$$N_2O_2(g) + O_2(g) + 2NO_2(g)$$

In aqueous solution, the NO_2 formed subsequently disproportionates to nitric and nitrous acids [18]:

$$2NO_2(aq) + H_2O \rightarrow NO_3^- (aq) + NO_2^- (aq) + 2H^+$$

$$k = 6.5 \times 10^7 \; M^{-1}s^{-1}, \qquad pK_a(HONO) = 3.35$$

or it can combine with NO to form nitrous acid or nitrites:

$$NO(g) + NO_2(g) + H_2O(g) \rightleftharpoons 2HNO_2(g)$$

$\Delta G° = -5.1$ kJ/mol, $\Delta H° = 38.2$ kJ/mol, $\Delta S° = 145$ J/mol, K, K $= 7.9$ atm [8]. The reverse of this reaction and the following provide pathways to evolve NO from NO_2^- in modest quantities at neutral pH [22].

$$3NO_2^- + H_2O \rightleftharpoons 2NO(aq) + NO_3^- + 2OH^-$$

$$K° = 9.1 \times 10^{-21}$$

In an aerated aqueous environment, the above reactions afford numerous ways in which NO can be rapidly destroyed to form NO_2 and eventually nitrates and nitrates. However, in low concentrations in the absence of oxygen it may survive considerably longer. Table 1 indicates that the radical nature of NO causes it to serve as both a relatively strong oxidant and reductant. Consequently, it can be expected to undergo a electron-transfer reactions with a number of biological redox partners. At low concentrations, reactions yielding nitrite might be expected to be favored, while hypoxic environments should afford NO sufficient stability to favor its coordination to transition metal sites. Indeed, metal ion complexation is probably necessary to preserve and transport NO in the biological mileu. NO might also be preserved *in vivo* through formation of thiol adducts (RS–NO), which may be metal assisted.

Nitric oxide also reacts with superoxide to form the stable peroxynitrite anion, which, however, decomposes on protonation [23].

$$O_2^- + NO \rightarrow ONOO^-$$

$$ONOO^- + H^+ \rightleftharpoons ONOOH$$

$$pK_a = 6.6 \ (0°), \ 7.49 \ (37°)$$

$$ONOOH \rightarrow HO^· + NO_2^·$$

$$k = \frac{k_{HA}[H^+]}{[H^+] + K_a}, \quad k_{HA} = 0.65 \ s^{-1}$$

$$HO^· + NO_2^· \rightarrow NO_3^- + H^+$$

$$k = \sim 3 \times 10^4 \ M^{-1}s^{-1}$$

The half-life for the decomposition to yield $HO^·$ is 1.9 s under physiological conditions. The recombination rate in the final reaction is slower than most other reactions involving $HO^·$, so that generation of hydroxyl radical in this fashion may be important physiologically [23].

Nitric oxide can be scavenged as a stable, ESR-detectable free radical by a quinodimethane intermediate, 1,2-disopropylidene-3,5-cyclohexadiene, which can be generated by photolysis of 1,1,3,3-tetramethyl-2-indanone. This approach offers the possibility of detecting and possibly quantitating NO produced by biological systems [24]. Free NO in solution is not detectable by ESR.

In scrupulously anaerobic basic solutions, NO can combine directly with thiolates to form thionitrites (S-nitrosothiols) [25], which can decay by several modes to yield NO.

$$RS^- + \cdot NO \rightarrow RS-\overset{.}{N}O \overset{H^+}{\rightleftharpoons} RS-\overset{.}{N}OH$$

However, if oxygen is present, N_2O_4 is the more likely nitrosating agent. In acidic solution, nitrous acid rapidly nitrosylates thiols according to the rate law:

$$Rate = k[HONO][RSH][H^+]$$

where k is 4.5×10^2 $M^{-2}s^{-1}$ for cysteine [26].

3 Medical Use of Sodium Nitroprusside

Sodium nitroprusside, $Na_2[NO(CN)_5Fe] \cdot 2H_2O$, which has been marketed as Nipride™ [27, 28] and more recently as Nitropress™ [29], is often used to lower blood pressure in humans in a titrametric fashion. Its hypotensive effect is evident within seconds after infusion and the desired blood pressure is usually obtained within 1–2 minutes. The intensity of the hypotensive response is modulated by small changes in its infusion rate (maximum: 10μg/kg/min in 5% dextrose) and continues only as long as administered. Consequently, nitroprusside is useful in cases of emergency hypertension, heart attacks and surgery [27, 28]. For example, during open heart surgery, hypothermia ($\sim 18\,°C$), which is induced to slow metabolism, often causes intense vasoconstriction that can be reversed with nitroprusside [30]. In post operative patients with low cardiac output, or in patients with congestive heart failure, nitroprusside is used to reduce the cardiac workload by decreasing peripheral vascular resistance [31–33]. As a precautionary measure, if adequate lowering of blood pressure is not achieved after 10 min, the infusion is normally discontinued to obviate the risk of cyanide toxicity, which may occur from cyanide released from the drug [27, 28, 34].

Cyanide poisoning poses some risk; however, this is minimized both by the kinetic inertness of both Fe^{II} and Fe^{III} cyano complexes and the high affinity of these ions for cyanide ($[Fe^{II}(CN)_6]^{4-}$, $\beta_6 \sim 10^{35}$; $[Fe^{III}(CN)_6]^{3-}$, $\beta_6 \sim 10^{43}$; [7, 35]. Small amounts of cyanide, which are released by photolysis and reduction products of nitroprusside, can usually be metabolized in the liver and kidneys by the enzyme rhodanase, which converts CN^- to SCN^- [7, 36]. Cyanide can also be taken up by hydroxocobalamin to generate cyanocobalamin (B_{12}). As the conversion of cyanide to thiocyanate is dependent on the availability of sulfur, thiosulfate can be administered as an antidote [37]. Monitoring thiocyanate levels as an indicator of cyanide toxicity is no longer routine, but is done on patients with severe hepatic compromise who have been maintained on nitroprusside for a prolonged period [31]. A major disadvantage of the drug is its photoreactivity, which necessitates its being stored in the dark and intravenous solutions to be shielded from light [27, 28].

Other nitrosyl compounds such as, $K[NOBr_5Ir]$, $K_2[NOCL_5Ru]$ and $[NO(NH_3)_5Ru]Cl_3$, have also shown vasodilatory activity, presumably by releasing NO, but are too toxic for clinical use. A nitro complex, $[Cl(O_2N)(bpy)_2Ru]$, which probably releases NO_2^-, exhibited higher activity than sodium nitrite [38].

4 Biological Functions of Nitric Oxide

4.1 Smooth Muscle Relaxation and Vasodilator Functions of Nitric Oxide

Nitroprusside's therapeutic properties depend upon its release of nitric oxide in the appropriate biological setting [7]. A long-sought natural vasodilator known as the Endothelial Derived Relaxing Factor (EDRF) [39], which plays an important role in the regulation of blood flow and is released by the endothelial cells lining blood vessels is now thought to be NO [21]. Both NO and EDRF were found to: 1) have short half-lives, 2) be inhibited by hemoglobin or hemeproteins [40], 3) be deactivated by superoxide, and 4) be strong activators of the enzyme, guanylate cyclase, which converts GTP to cGMP. Consequently, the physiological effects of NO are linked to those of cGMP [41]. The vasodilator action of nitroprusside appears to derive from its acting as a messenger to stimulate cGMP production and so reversibly decrease voltage activated Ca^{2+} channel currents across the plasma membrane and the sarcoplasmic reticulum in smooth muscle cells that constrict arterial walls [42]. cGMP can act directly on ion channels and activates protein kinases, which catalyze the phosphorylation or dephosphorylation of proteins [43]. cGMP directly regulates membrane cation channels in retinal photoreceptor cells, mediates smooth muscle relaxation, inhibits platelet aggregation (anithrombic effect), modulates excretion of Na^+ by the kidney, and helps to regulate cardiac function by modulating Ca^{2+} currents [13, 41]. Nitroglycerine, sodium azide, hydroxylamine, and sodium nitrite are also believed to relax smooth muscle by increasing cGMP levels through serving as substrates for the generation of nitric oxide [44, 45]. Both NO and N_3^- produce violent convulsions upon penetrating the cerebroventricular system of the brain [46].

While it is clear that NO is involved in the EDRF, there is also evidence that conjugation with thiols may occur as intermediates [47]. In 1981, Ignarro observed that NO-donating vasodilators react with cysteine to form S-nitrosocysteine, an activator of guanylate cyclase, and suggested that the formation of unstable S-nitrosothiols was involved in the biological function of nitric oxide

[48, 49]. Indeed, vasodilators require cysteine or thiols to activate guanylate cyclase and S-nitrosothiols are the products of the reaction of thiols and NO. Moreover, S-nitrosothiols have the same therapeutic properties as NO and the instability of S-nitrosothiols is consistent with the short duration of nitroprusside's hypotensive action [48, 50 51]. Nitrosothiols can release NO through several modes of decay: 1) homolytic scission of the RS:NO bond, 2) displacement by a thiyl radical to generate a disulfide, and 3) single electron reduction to the thiol(ate) [52]. On the other hand, other reductants also stimulate the nitroprusside activation of guanylate cyclase and both RS:NO formation and decomposition are probably slower than the reaction of NO with O_2 [53, 54]. It may be that as NO is generated, either by NO-synthase or from a metallonitrosyl, it is delivered to guanylate cyclase both by diffusion and through a slower, but somewhat longer-lasting, thiol-mediated path, which assumes importance in environments low in free O_2. Formation of thionitrite adducts from nitroprusside is also possible (see below).

Nitric oxide appears to be the modulator of neuronal activity released by NANC (non-adrenergic, non-cholinergic) nerves in the peripheral nervous system [41]. In functioning as a NANC neurotransmitter, the physical and chemical properties of NO coupled with its ability to open blood vessels allow it to play a role as a paracrine mediator in penile erection. Its small, lipophilic character allows it to readily and rapidly diffuse out of cells where it is synthesized and into adjacent target cells. To be effective, smooth muscle cells must function as a group. The rapid and profound vascular smooth muscle relaxation required to achieve and maintain penile erection is a good example of how NO can coordinate the actions of neighboring cells. In fact, NO-donating vasodilators such as nitroprusside and nitroglycerine have long been used to treat impotence [1] and NO-synthase inhibitors can block this physiological response [55]. A possible mechanism for nitric oxide's role is that electrical field stimulation of NANC neurons causes an influx of Ca^{2+}, which activates NO-synthase. The NO then diffuses into the corpus cavernosum to activate guanylate cyclase to produce cGMP, which induces smooth muscle relaxation in this region [1].

NO is also a likely candidate for the NANC messenger of the myenteric plexus of neurons in the gastrointestinal tract, which mediate peristaltic movements. These neurons are rich in NO synthase and inhibitors of this enzyme prevent the nerve-evoked relaxation of the gut [12]. Nitric oxide released from nitroprusside also stimulates ADP ribosylation of glyceradehyde 3-phosphate dehydrogenase [56]. Consequently, there may be a number of enzymes, which are probably Fe-centered, that may be activated by NO.

4.2 Messenger Role of NO in the Central Nervous System

Endogenous nitric oxide plays an important role as a messenger between neurons at synapses in the central nervous system [11] and fulfills the major criteria for a neurotransmitter [12]. Not all neurons produce and release NO, which is associated with the N-methyl-D-aspartate (NMDA, a glutamate analog) receptor, whose function is to trigger long-lasting changes in synaptic strength and to generate activity-dependent organization of some nerve fibers during development. Both processes involve a retrosignal from the postsynaptic to the presynaptic neurons. As with smooth-muscle-cell coordination, NO acts as a diffusible, short-lived [57] molecule within a local space to help shape the three-dimensional network of synaptic responses [41].

In the cerebellum, which coordinates motor functions in a well-defined synaptic network, NO synthase occurs in basket cells and their horizontal processes that synapse on Purkinje cells, and in mossy fibers that synapse on granule cells [58] and has also been reported in the granule cells [41]. Nitric oxide concentrations in cerebellar slices have been measured to be about 50 nM following repeated electrical stimulation of afferent fiber pathways [20]. The Purkinje cells, which are the final common output path of cerebellar activity, contain high concentrations of guanylate cyclase and cGMP-dependent protein kinase [58]. Consequently, the Purkinje cells and, possibly, the Bergmann glial cells are two potential NO-targets, which are in appropriate locations relative to the post-synaptic membrane [41]. In the cerebellum coincident stimulation of the parallel and climbing fibers leads to depression of the ability of the parallel fibers to activate the Purkinje cells. This phenomenon, which is termed long term depression (LTD), bears some analogies to long term potentiation (LTP) described below for the hippocampus and may also involve NO as a modulator [58]. In fact, nitroprusside administration can elicit this phenomenon.

It now appears that NO, functioning as a neuromodulator in the hippocampus of the brain plays a fundamental role in the basic neuronal processes initiating learning and experience [13, 59]. As these abilities are among the very essences defining humanity, understanding them at the molecular level is of fundamental importance. Memory begins with an event that potentiates a neuronal synapse or (more likely) a network of synapses to fire again upon the same or a similar stimulus. Since this potentiation must be relatively long-lasting, even for short-term memory, this process is called long-term potentiation. The hippocampus is associated with short term memory, and those with a damaged hippocampus cannot form new memories nor, consequently, learn from new experience.

The mechanism of LTP in the hippocampus appears to involve the secondary neuromodulators Ca^{2+} and NO. The LTP phenomenon exhibits distinctive properties in that: 1) It is specific to activated synapses. 2) It is cooperative, showing a stimulus threshold before developing and usually involves simultaneous activation by many inputs. 3) It is associative in requiring simultaneous pre- and postsynaptic activity. The initiating step occurs at the postsynaptic neuron and the maintenance step occurs, at least in part, at the presynaptic neuron [60]. Since learning tends to occur in situations that are either new or striking, or repetitive, there may be more than one mechanism for potentiating neural networks to fire (i. e. thoughts to occur) under the appropriate stimuli.

The hippocampus has three distinct pathways capable of LTP: 1) Perforant pathway, which involves axons from the subiculum synapse with the granule cells in the dentate gyrus. 2) Mossy fibre pathway, in which axons extending from the dentate gyrus synapse with the pyramidal cells in the CA3 region. 3) Schaeffer collaterals, which involve axons from the pyramidal cells in the CA3 region synapse with the pyramidal cells in the CA1 region. The effect of different compounds on LTP can be monitored by inserting an electrode into one of the excitatory input pathways and stimulating it at a high frequency (100 Hz) for a period of less than a second. An extracellular recording electrode measures the voltage response in the appropriate postsynaptic region. A modest stimulus generates a typical response. Following a strong stimulus, that induces LTP, a greater response is reproducibly generated by the same modest stimulus. The in vitro efficacy of a NO-releasing drug, such as nitroprusside, can be determined by the voltage response reproducibly generated after LTP induction in the presence of the drug [60].

Following transmission of the neural message as a voltage pulse mediated by a change in Na^+/K^+ concentration in the axon, the neuronal message must be transmitted form one nerve to the next through the synapse. This is usually accomplished by vesicular release of neurotransmitters. While several different neurotransmitters may be contained in a given vesicle, those released in the hippocampus usually contain a high concentration of glutamate. Upon vesicle release, the glutamate travels from the presynaptic neuron into the synapse and then arrives at (among others) two different types of receptors on the postsynaptic membrane. Since one of these receptors also binds the synthetic neurotransmitter, N-methyl-D-aspartate (NMDA), it is termed the NMDA receptor.

On binding glutamate, the non-NMDA (or AMPA) receptor opens ion channels that allow Na^+ to rush in, thereby depolarizing the membrane. This depolarization opens a second ion channel by reducing the Mg^{2+} block associated with the NMDA receptor, through which Ca^{2+} ions move. Calmodulin binds the incoming Ca^{2+} to trigger the LTP response, by activating the calmodulin-dependent enzyme, nitric oxide synthase, directly and indirectly through stimulating several Ca^{2+}-calmodulin dependent kinases (α-CaMKII, protein kinase C and tyrosine kinase) [61, 62]. Unlike the macrophage NOS, the brain and endothelial NO-synthases are highly dependent on calcium concentrations. At normal resting cystolic free $[Ca^{2+}]$ (~ 50 nM) the enzyme is inactive, but is maximally active at $[Ca^{2+}] = 0.4 - 1$ μM [41].

NO-synthase releases NO as it catalyzes the oxidation of arginine to citrulline [10, 63]. The NO diffuses out to neighboring astrocyte fibers and back to the presynaptic neuron, where it activates one of two enzymes, guanylate cyclase or ADP-ribosyltransferase, initiating biochemical changes that stimulate the additional release of the neurotransmitter glutamate [13, 41]. In mediating glutamate-stimulated cGMP production, NO may generate an allosteric effect by binding to iron in the heme portion of guanylate cyclase causing a conformational change in the protoheme and possibly untethering it from the protein [64, 65].

4.3 Nitric Oxide as a Macrophage Defense Weapon

Nitric oxide is produced by macrophages to kill both invasive cells (neoplasms, bacteria, fungi, etc.) [66] and in the rejection of tissue grafts [67]. When macrophages are activated by interferon-γ, endotoxins (bacterial cell wall lipopolysaccharides that elicit inflammatory responses), or T cells, they respond by converting arginine into NO, which can be observed as nitrite or nitrate in the tissue [11] or by EPR detection of NO-hemoglobin and other Fe^{II}-NO· complexes [67–69]. The tumor-antagonizing ability of macrophages disappears when arginine is removed from the medium. In murine macrophages the induced $[NO·]_{aq}$ has been measured at a steady-state level of 0.43 μM [66]. The toxic effect appears to derive from inhibition of enzyme activity through nitric oxide binding to nonheme iron-centers in proteins, such as aconitase and ribonucleotide reductase, and release of intracellular iron from the cells targeted by the macrophages [41, 70, 71].

Both inducible and constitutive forms of NO-synthase exist in macrophages. Both are: 1) cystolic dioxygenases that incorporate oxygen into both NO and citrulline [72], 2) NADPH dependent, and 3) inhibited by L-arginine analogs. However, the constitutive enzyme is calmodulin dependent and generates picomolar quantities with short-lasting release, while the induced form generates nanomoles with long-lasting release and its induction can be inhibited by glucocorticoids [11]. Macrophages that are stimulated to produce NO by interferon-γ or lipopolysaccharide can be inhibited with N-iminoethyl-L-ornithine (L-NIO). In some cells the induced form appears to be dependent on tetrahydrobiopterin and is stimulated by flavin adenine dinucleotide and reduced glutathione [73].

4.4 Nitric Oxide Toxicity

The essential role played by NO in the nervous and vascular systems is in sharp contrast to the high degree of toxicity associated with its radical character. In fact, its rapid reactivity with oxygen and other substances in water suggest that, in some instances, NO transport may be effected by metal ion complexation. Mechanisms of NO toxicity include: 1) Coordination to or oxidation of the Fe^{II} in hemoglobin to generate methemoglobin thereby decreasing the oxygen carrying capacity of the blood [46]; 2) Binding to ferrous heme and iron-sulfur centers of enzymes, such as those involved in the electron transport chain, citric acid cycle and DNA synthesis [7, 41]; 3) Hydroxyl radical formation. NO reacts with superoxide to form the peroxynitrite anion (ONOO–), which decomposes to yield highly damaging $\cdot OH$ and $\cdot NO_2$. Superoxide dismutase protects against this mode of toxicity, since the half-life of NO in the blood is approximately doubled when this enzyme is present [23, 41]. 4) Neurotoxicity arising from excessive concentrations of NO as a neuromodulator. If placed directly in the brain, azide and nitrite produce seizures at low levels by rapid conversion to nitric oxide [46].

Hemoglobin can serve as a buffer against NO toxicity, since it coordinates NO and NO-precursors such as hydroxylamine and nitrite. Consequently, these inorganic vasodilating agents usually fail to penetrate the blood-brain barrier protecting the CNS. The affinity between NO and hemoglobin may be one reason why erythrocytes are excluded from neuronal spaces. Excessive NMDA receptor activation and subsequent Ca^{2+} entry and/or superoxide generation may contribute to neurodegenerative conditions such as chronic epilepsy and Alzheimer's disease [13, 41]. Indeed, NMDA stimulation of NO-producing cells, which have extensive ramification into surrounding cells, results in clumps of dead neurons adjacent to these ramifications. In the diseases mentioned, NO-producing cells seem to be able to survive in these disease and may have a protective mechanism against NO toxicity that other cells lack [12]. Nitric oxide may also be released as a result of brain artery damage occurring in strokes, which results in neurotoxic effects [12].

Damage arising from peroxynitrite formation might be expected to be more prevalent in the neuronal mitochondria, where superoxide is sometimes released by cytochrome oxidase, and increase with age as activity levels of superoxide dismutase decline. Subsequently formed hydroxyl radicals would attack enzymes including cytochrome oxidase, conceivably altering this protein to allowing the escape of additional superoxide. Moreover, generation of $HO\cdot$ can also lead to lesions in the mitochondrial DNA that could mutate or affect the production of mitochondrial proteins. Again, cytochrome oxidase may be of particular interest, since it decreases in some aged-linked disorders like Parkinson's disease [74].

5 Reaction Chemistry of Nitroprusside and Related Metallonitrosyls

5.1 Overview of Reactions of Nitroprusside

Sodium nitroprusside is the only clinically used metal complex of NO, so that its reactions provide an indication of the types of reactivity that metallonitrosyl complexes might be expected to have in physiological environments (see Fig. 1). The in vivo activation of nitroprusside depends on its reduction to $[Fe(CN)_5NO]^{3-}$, which then releases cyanide to give $[Fe(CN)_4NO]^{2-}$ which in turn releases NO and additional CN^- to yield $Fe_{(aq)}^{II}$ and $[Fe(CN)_6]^{4-}$ [75]. $[Fe(CN)_5(NO)]^{3-}$ is paramagnetic ($g_x = 1.9993$, $g_y = 1.9282$, $g_z = 2.008$,

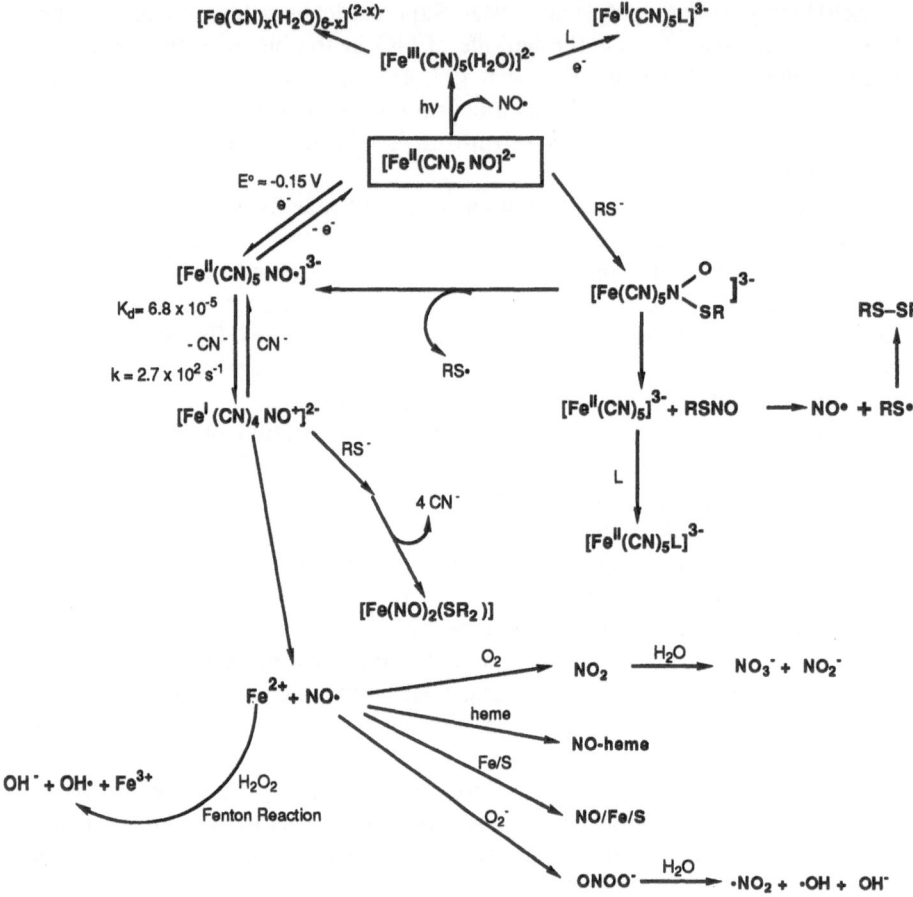

Fig. 1. Potential metabolic reactions of nitroprusside

$\lambda_{\max} = 405$ nm, $\nu_{NO} = 1580$ cm^{-1}) [76, 77] with the unpaired electron localized on the NO [78]. By way of comparison, similar chemistry has been noted for the one-electron reduction product of isoelectronic nitrosyl complexes, such as [NO(NH$_3$)$_5$Ru]$^{3+}$ [79] and NO(bpy)$_2$ClRu]$^{3+}$ [80], in which the added electron is also localized mainly on the nitrosyl. Like its precursor, [Fe(CN)$_4$(NO)]$^{2-}$ is paramagnetic ($g_x = 2.036$, $g_y = 2.0325$, $g_z = 2.0054$), but with a different absorption maximum ($\lambda_{\max} = 696$ nm) and a significantly higher nitrosyl stretching frequency ($\nu_{NO} = 1755$ cm^{-1}) [76, 77]. In contrast to its precursor, ESR and IR measurements indicate that the odd electron in [Fe(CN)$_4$(NO)]$^{2-}$ is localized more on the iron than on the nitrosyl [75, 78]. Since this places the unpaired electron in an antibonding iron orbital, [Fe(CN)$_4$(NO)]$^{2-}$ is substitution labile.

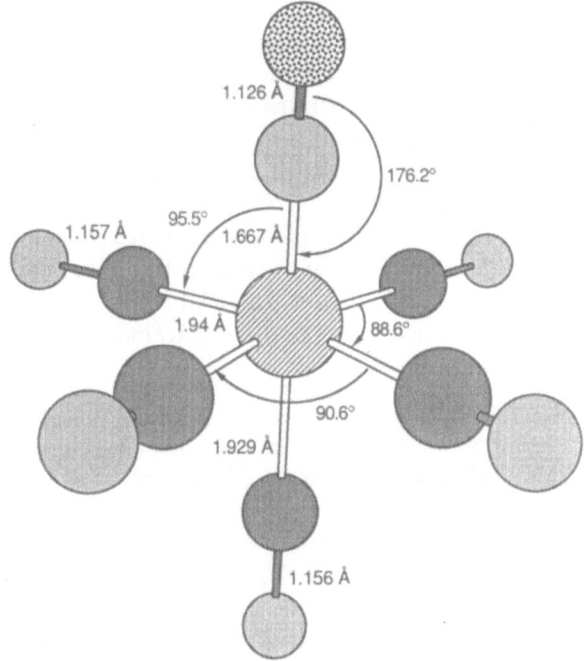

Fig. 2. Neutron diffraction structure of the nitroprusside ion in $Na_2[Fe(CN)_5NO]·H_2O$ [83]

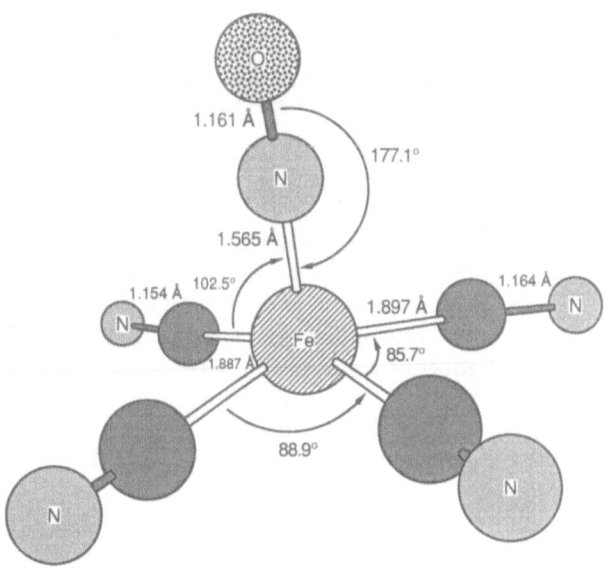

Fig. 3. X-ray diffraction structure of $[Fe(CN)_4NO]^{2-}$ from $(Et_4N)_2[Fe(CN)_4NO]$ [76]

The structures of the nitroprusside ion [81–83] and $[Fe(CN)_4NO]^{2-}$ are shown in Figs. 2 and 3, respectively. The geometry of the nitroprusside ion is typical for a low-spin complex with a Fe^{II}–NO^+ core and stronger π-backbonding to the NO relative to the *trans* cyanide causes a slight displacement (0.18 Å) of the Fe toward the nitrosyl. In $[Fe(CN)_4NO]^{2-}$, π-bonding to the nitrosyl pulls the Fe^{II} about 0.41 Å out of the plane of the four cyanide ligands [76]. Curiously, the short Fe–NO bond distance and the linear, rather than bent, Fe–N–O core are unusual for a square pyramidal, d^7, complex and are closer to those seen in d^6, Fe^{II}–NO^+ species. The structures of $Na_2[NO(CN)_5M]\cdot 2H_2O$ (where M = Fe [82] and Ru [84]) are isotypic (*Pnmm*), with similar nitrosyl bond distances and angles (Ru–N, 1.773(3) Å; N–O, 1.130) 4) Å and Ru–N–O, 174.4°(3), cf. Fig. 2).

5.2 Redox Reactions of Nitroprusside

Detailed polarographic studies indicate that nitroprusside exhibits three reduction waves: -0.123 V, -0.351 V, and -1.08 V (vs. N.H.E.) [85, 86]. In acetonitrile/water, these occur at -0.24 V, -0.56 V and -1.65 V, which suggests that these potentials may have a marked media-dependence. The first arises from a pseudoreversible one-electron transfer (in the presence of CN^-) to form $[Fe(CN)_5(NO)]^{3-}$, which otherwise rapidly loses CN^-, and the second from the one-electron reduction of the resulting $[Fe(CN)_4(NO)]^{2-}$ [86], which is reversible on a rapid time scale. The diminution of the second wave by half at a nitroprusside concentration of 10 mM suggests dimerization to $[NO(CN)_4FeCNFe(CN)_4(NO)]^{5-}$ [86, 88]. The third wave is an irreversible multi-electron process [3, 87]. In water, the first two couples exhibit an ionic strength dependence of **84** $mV/\mu^{1/2}$ and **145** $mV/\mu^{1/2}$, respectively [88]. The one-electron reduced form can also be generated chemically by reduction with borohydride, ascorbate, dithionite or superoxide [7] and is reversibly oxidized by O_2 [43, 86]. The reduction products from the multi-electron wave lose CN^- more rapidly than those from the first two waves, resulting in a mixture of products of the general type $[Fe(CN)_{5-x}(H_2O)_x(NO)]^{n-}$.

Hydrated electrons generated by pulse radiolysis reduce nitroprusside at diffusion limited rates to $[Fe(CN)_5(NO)]^{3-}$, which undergoes unimolecular decay within milliseconds to yield $[Fe(CN)_4(NO)]^{2-}$, which, in turn, persists for minutes in aqueous solution [89, 90].

$$k_f = 2.7 \times 10^2 \, s^{-1}$$

$$[Fe(CN)_5(NO)]^{3-} \rightleftharpoons [Fe(CN)_4(NO)]^{2-} + CN^-$$

$$k_f = 4 \times 10^6 \, M^{-1} s^{-1}$$

$$K = 6.8 \times 10^{-5}$$

The first reduction product, $[Fe(CN)_5(NO)]^{3-}$, exhibits a $pK_a \sim 5.8$. Organic radicals combine with $[Fe(CN)_5(NO)]^{3-}$ near the diffusion limit to produce $[Fe(CN)_5N(O)R]^{3-}$ [89, 90]. The blue $[Fe(CN)_4(NO)]^{2-}$ can be voltammetrically oxidized ($E° = -0.42$ V) and reduced ($E° = -0.99$ V) in one-electron steps, which are reversible on the cyclic-voltammetric time scale [87].

In erythrocytes, nitroprusside is probably reduced by hemoglobin and methemoglobin reductase. Hepatocytes reduce nitroprusside more efficiently through a NADPH: cyt-P450 reductase in the microsomes, while in the mitochondria dehydrogenases are most effective at this function. Because CN^- liberated following reduction may block cellular respiration, mitochondrial reduction may be important in nitroprusside toxicity. Also, $[Fe(CN)_5(NO)]^{3-}$ can be re-oxidized by oxygen to produce O_2^- and H_2O_2, which can then be further reduced through nitroprusside redox cycling to produce the highly toxic $HO\cdot$. While a number of thiols are capable of reducing nitroprusside, they are less efficient than enzymatic processes, but may contribute to cyanide toxicity by displacing cyano ligands from reduced nitroprusside products. Para-hydroquinones, catechols and catecholamines can also reduce nitroprusside, and such reduction may be important in the kidney where there is a high concentration of catecholamines in the chromaffin cells [43].

Thiolates reduce nitroprusside stoichiometrically to yield R_2S_2 and $[Fe(CN)_5(NO)]^{3-}$, which can then convert to $[Fe(CN)_4(NO)]^{2-}$ with subsequent release of NO. Reduction may take place through thionitrito adduct formation (see below) followed by generation of the redox products and eventually $[Fe(CN)_6]^{4-}$ or $[Fe(CN)_5L]^{n-}$ or release of the thionitrite ligand with subsequent decay to the same products [7, 54]. If oxygen is present, nitroprusside is regenerated and can catalyze the air oxidation of thiols, such as cysteine to cystine [91].

Nitroprusside is capable of ligating redox partners through a cyano ligand to participate in inner-sphere electron-transfer reactions. With hemoglobin an electron is transferred in this fashion from the porphyrin center to yield $[Fe(CN)_4(NO)]^{2-}$, methemoglobin and NO (see below) [92].

5.3 Correlations of Metallonitrosyl Properties with the Lever Parameter

The ability of NO to function as both a π-donor and π-acceptor ligand leads to a number of well known structural differences in metallonitrosyl complexes with regard to the M–N–O bond distances and angles [93, 94]. The nitrosyl's capacity to accept π-electron density from the metal is dependent on the inclination of the metal to backbond, which is (in part) a function of the π-acidity of the other ligands. The net π-donor (acceptance) of the other ligands can be quantified by the sum of the Lever electrochemical parameters (ΣE_L) for the nonnitrosyl ligands. The Lever parameter defines each ligand's place in an eletrochemical series such that the sum of these parameters correlates well with the reduction potential for a given complex [95]:

$$E^° = S_M \sum_{i=1}^{n} a_i E_{L(i)} + I_M$$

where $E^°$ is the reduction potential of the metal ion couple, n is the coordination number and a_i is the dentacity of a given ligand.

The tendency of the nitrosyl, rather than the metal ion, to accept an added electron and the noninnocent nature of NO, which arises from its ability to exist in oxidation states between NO^- and NO^+, has lead to estimates for its E_L which range between 1.3 and 2.5 [96, 97]. Consequently, the E_L for "NO" can

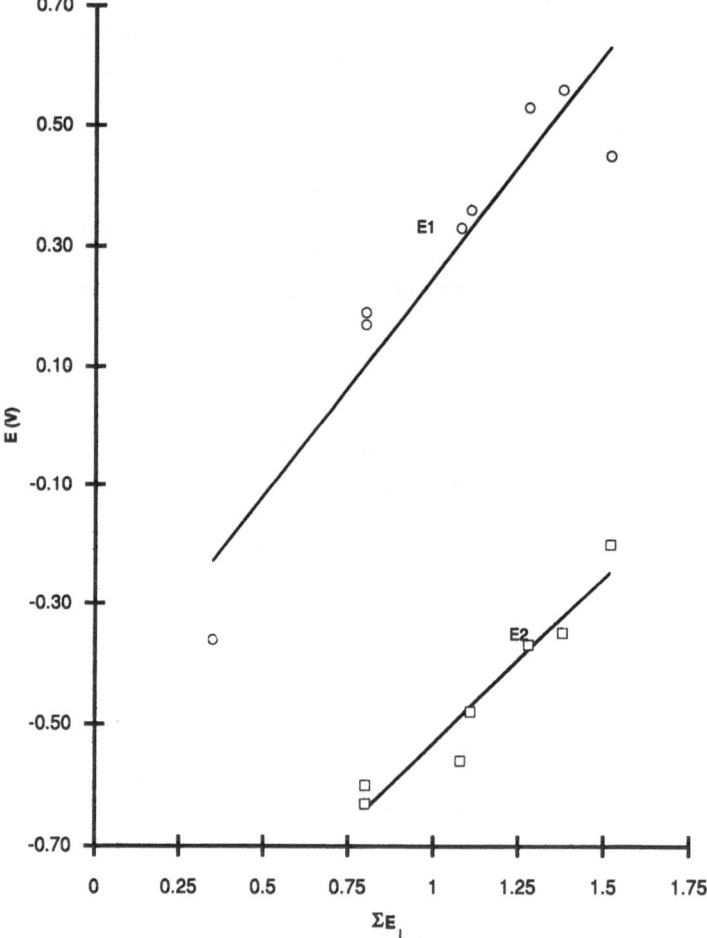

Fig. 4. Correlation between coordinated nitrosyl reduction potentials on Ru^{II} amine complexes and ΣE_L. For E_1 ($NO^+ + e \rightarrow NO\cdot$), slope = 0.734 V/E_L, intercept = -0.485 V. For E_2 ($NO\cdot + e \rightarrow NO^-$), slope = 0.546 V/E_L, intercept -1.708 V. Data from Refs [98] and [99]

be expected to vary with both the metal ion and the other ligands. Indeed, since electrons added to low-spin d^6 metal ions, such as nitroprusside and ruthenium(II) amine complexes, often enter a π^* nitrosyl-centered orbital [98], the reduction potentials for coordinated nitrosyls are a linear function of ΣE_L for the other ligands (Fig. 4). Since nitroprusside is activated by reduction to release NO in the body, the ability to adjust the nitrosyl reduction potential is a key factor in the design of metallonitrosyl pharmaceuticals. A second important parameter is how rapidly this electron can be shunted from the π^*-NO orbital into an antibonding σ^* (d_{z2} or d_{x2-y2}) orbital so as to weaken the M-NO bond and release NO.

Other useful correlations exists between ΣE_L and the ability of the metal to backbond to the nitrosyl as reflected in the Tc–NX (X = O, S) bond distance (see Fig. 5 [96] and the NO stretching frequency in Ru^{II} amine complexes (Fig. 6). As ΣE_L increases, the ability of the metal to backbond decreases and the coordinated NO becomes more like NO^+. Such correlations should prove useful not only for predicting the ability of metallonitrosyls to release NO, but also the tendency of the nitrosyl to undergo nucleophilic attack.

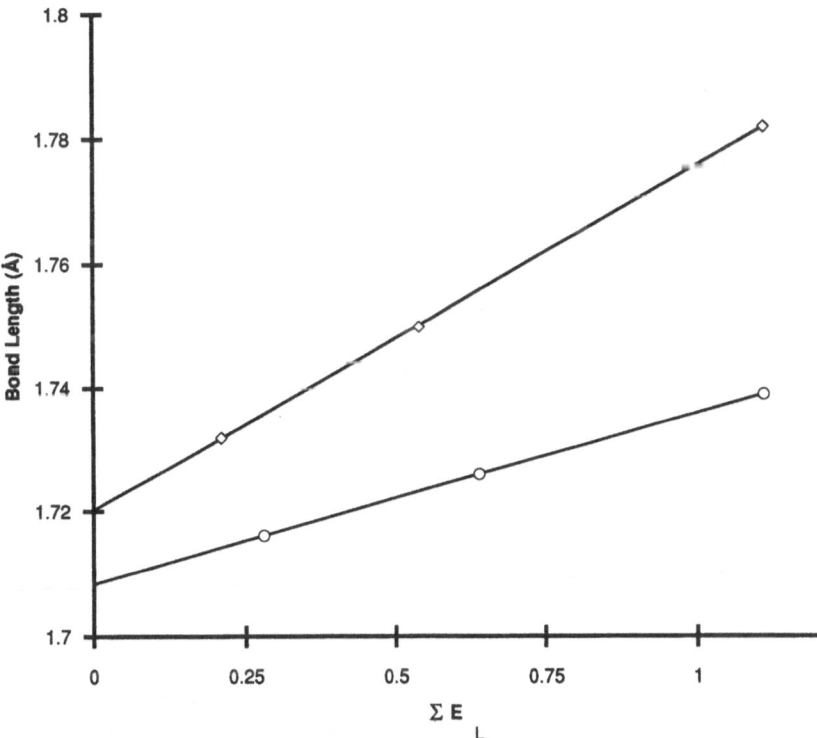

Fig. 5. Correlation between Tc-NX$^+$ bond lengths and ΣE_L. *Circles*, X = S; *squares*, X = O. Slopes and intercepts, respectively, are 0.0277(1)Å/E_L and 1.708(1)Å for the nitrosyl and 0.0556(2)and 1.720(1)Åfor the thionitrosyl complexes [96]

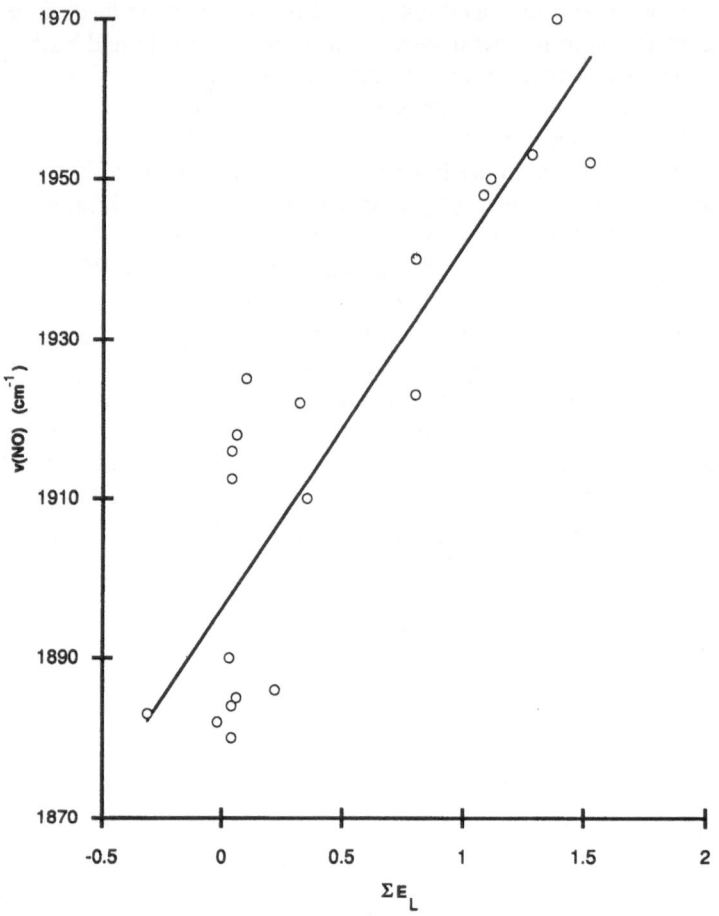

Fig. 6. Correlation between ν_{NO} and ΣE_L for amine complexes of Ru^{II}. Slope = 45 cm^{-1}/E$_L$, intercept = 1896 cm^{-1}. Data from Refs. [98], [100] and [101]

5.4 Reactions of Nitroprusside with Hemoproteins

Conflicting studies exist on the effect of blood on the stability of nitroprusside. One report indicates that in the dark there is no increased cyanide concentration in blood samples after four hours [102]; however, other recent work indicates that normal light has little effect on nitroprusside in phosphate buffer at pH 7 and that erythrocytes do generate CN⁻ from nitroprusside, since lysates of these cells quantitatively release all the cyano ligands within 3 hours [103]. Nitroprusside can react with oxyhemoglobin to oxidize it through an inner-sphere (cyano-bridged) mechanism to methemoglobin [92, 103], which then binds free cyanide to produce cyanomethemoglobin. Following reduction of the nitroprusside, cyanide is released making it available for coordination to Fe^{II-}

heme and $[Fe(CN)_4NO]^{2-}$ is formed. Consequently, the reduced complex, $[Fe(CN)_4NO]^{2-}$, appears to be the nitrosyl donor, which reacts with a second hemoglobin molecule to yield nitrosylhemoglobin [104]. Glutathione in the blood can reduce both nitroprusside and methemoglobin, thereby avoiding high levels of the latter [103]. High cyanide concentrations can inhibit nitrosylation of hemoglobin by lowering the concentration of $[Fe(CN)_4NO]^{2-}$ presumably by shifting the equilibrium toward $[Fe(CN)_5NO]^{3-}$ ($K_d = 6.8 \times 10^{-5}$) [7, 36, 104].

While it has been implied that nitrosyl transfer between metal sites can occur via μ-bridged nitrosyl intermediates [105], which are fairly common in organometallic nitrosyl complexes [106], experimental evidence for this in aqueous solution has been based on product analysis [107] rather than direct observation. Coordinatively unsaturated complexes such as $(NO)Co^{II}(DMG)_2$ (DMG = dimethylglyoxime) can bind and transfer NO to the heme iron in hemoglobin, but they do so by release of the NO following coordination to amino acid functional groups such as ß93-cys [107]. The nitrosyl group can be cyclically transferred between cobalt (II) and iron porphyrins as a function of the oxidation state of the iron [108]. While it would appear difficult for most metal complexes to penetrate into the heme cavity of proteins to form μ–NO intermediates, some small metal complexes may be capable of directly transfering NO to a heme near the protein surface.

5.5 Photochemistry of Nitroprusside

Nitric oxide can be released by photodegradation of nitroprusside, which occurs in high quantum yield by light of 366 nm-436 nm [102, 109, 110]. The product is $[Fe^{III}(CN)_5(H_2O)]^{2-}$, which is substitution-liable and possesses no hypotensive properties [75, 111]. In a variety of other solvents, $[Fe^{III}(CN)_5S]^{2-}$ (S = solvent) is formed through what is probably a dissociative/cage recombination mechanism [109, 110]. A minor reaction pathway, which increases in highly nucleophilic solvents, leads to the formation of NO^+ and $[Fe^{II}(CN)_5(H_2O)]^{3-}$ [109]. Irradiation of a neutral solution of nitroprusside anion results in an initial pH decrease, which is probably due to the production of nitrous acid from the aerobic aqueous reaction of the NO released [112]. Nitroprusside ion can also be photoreduced in aqueous solution to yield $[Fe(CN)_5NO]^{3-}$ [109]. When nitroprusside is irradiated with moderate intensity light in water over long periods of time, $[Fe(CN)_5NOH]^{2-}$, is also observed.

The electronic spectrum of nitroprusside with band assignments is given in Table 2. Irradiating the band at 321 nm results in nonspecific decomposition with ferric ion production. Photolysis with light of wavelength greater than 300 nm produced Prussian blue. Irradiation of the band near 400 nm excites an electron from a bonding metal-centered orbital to an antibonding one localized on the nitrosyl, which can result in electron emission or release of NO [112].

Nitroprusside has two metastable states (MS) that lie 9000 cm^{-1} (MSI) and 6500 cm^{-1} (MSII) above the ground state [113]. MSI is accessed by light of

Table 2. Aqueous electric spectra of relevant metallonitrosyl complexes, their reduction products, and reduction potentials [79, 89, 98, 112]

Ion	Transition	λ (nm)	ε (M^{-1} cm^{-1})	E° (V)
[Fe(CN)$_5$(NO)]$^{2-}$	$d_{xy} \rightarrow \pi^*$CN	200	24,000	-0.12
	$d_{xz,yz} \rightarrow d_{x2-y2}$ (sh)	238	700	
	$d_{xz\,yz} \rightarrow d_{z2}$ (sh)	265	900	
	$d_{xy}' \rightarrow d_{x2-y2}$ (sh)	321	47	
	$d_{xz,yz} \rightarrow \pi^*$(NO)	398	23	
	$d_{xy} \rightarrow \pi^*$(NO)	510	7	
[Fe(CN)$_5$(NO)]$^{3-}$		345	3,500	-1.08
		430	550	
[Fe(CN)$_4$(NO)]$^{2-}$		350	300	-0.35
		430	100	
		615	380	
[Ru(NH$_3$)$_5$(NO)]$^{3+}$		304	69	-0.12
		458	13.9	
[Ru(NH$_3$)$_5$(NO)]$^{2+}$		280	3,850	
		360	995	
[Ru(bpy)$_2$Cl(NO)]$^{2+}$	$d_{xz,yz} \rightarrow \pi^*$(NO)	336	10,000	0.44
	$d_\pi \rightarrow \pi^*$(bpy)	449	7,800	

380–550 nm and MSII by light between 380 nm and 460 nm. Below 100 K, they are long lived ($\tau > 10^7$ s), but thermally relax above 165 K. Both metastable states involve a perturbation of the N ≡ O bond as ν_{NO} is lowered from 1950 cm^{-1} to 1832 cm^{-1} in MSI and to 1664 cm^{-1} in MSII. While the transitions giving rise to these MS have been considered to be metal to ligand charge transfers (d → π^*), their extreme longevity at low temperature suggest a high Franck-Condon barrier between the GS and MS. Güdel has suggested that this could be achieved if the configurations of the GS and MS differ by two electrons such that the excited state is $t_{2g}^4\pi^*$(NO)^2 [113].

5.6 Substitution Reactions on Nitroprusside

While nitroprusside itself is substitution inert, the reduction product, [Fe(CN)$_5$(NO)]$^{3-}$, rapidly loses the *trans* cyanide (see above) and [Fe(CN)$_4$(NO)]$^{2-}$ is substitution labile via a dissociative pathway and is rapidly substituted by chelating diamines to give products such as [Fe(bpy)(CN)$_3$(NO)]$^-$, [110, 114]. If excess cyanide is present, [Fe(CN)$_6$]$^{4-}$ is formed [7]. In the presence of thiolates [Fe(NO)]$_2$(SR)$_2$]$^-$ and then [Fe$_2$NO$_4$(SR)$_2$]$^-$ is formed, while with HS$^-$ [Fe(NO)$_2$(SH)$_2$]$^-$ and then the tetranuclear [Fe$_4$S$_3$(NO)$_7$]$^-$ occurs [115].

5.7 Reactions of Coordinated Nitric Oxide

Nitrosyl complexes, in which $v_{NO} > 1886$ cm^{-1} or $F_{NO} > 13.8$ mdyn/Å, usually react as electrophilic nitrosating agents so that the ligand can be considered NO^+ [26]. Nucleophilic attack on the nitrosyl nitrogen is a common reaction encountered in the chemistry of nitroprusside and the rates and activation parameters for a number of different nucleophiles are listed in Table 3. Hydroxylamine adducts to nitroprusside via a rate law that is first order in the complex, the ligand and hydroxide ($k = 4.5 \times 10^5 M^{-2} s^{-1}$).

Nitroprusside reacts with hydroxide to yield the corresponding nitro complex. The rate is first order in both reactants and probably involves nucleophilic attack by OH^- on the NO followed by proton ionization. Infrared studies on ^{18}O-labeled compounds indicate the NO_2^- to be bound as nitro ($-NO_2$) rather than as nitrito ($-ONO$) [3, 116].

$$[Fe(CN)_5NO]^{2-} + 2OH^- \rightleftharpoons [Fe(CN)_5(NO_2)]^{4-} + H_2O$$

$$K = 1.0 \times 10^6, \quad \Delta H^\circ = -67.8 \text{ kJ/mol}, \quad \Delta S^\circ = -109 \text{ J/mol K}$$

For comparison [3, 117]:

$$NO^+ + 2OH^- \rightleftharpoons NO_2^- + H_2O$$

$$K = 2.3 \times 10^{31}$$

$$NO^+ + H_2O \rightleftharpoons HNO_2 + H^+$$

$$K = 5.0 \times 10^6 \ (20\,^\circ C)$$

Consequently, it is apparent that in nitroprusside NO^+ is highly stabilized by the metal ion against conversion to NO_2^-. Nitrite is then lost from $[Fe(CN)_5NO_2]^{4-}$ to yield $[Fe(CN)_5H_2O]^{3-}$ ($K = 3.0 \times 10^{-4}$, $\Delta H^\circ = 51.5$

Table 3. Rates and activation parameters for nitroprusside adduct formation with nucleophiles
$X^{n-} + [Fe(CN)_5NO]^{2-} \rightarrow [Fe(CN)_5N(O)X]^{(2+n)-}$

Nucleophile	k (25 °C) M^{-1}S^{-1}	$\Delta H^\ddagger$ kJ/mol	$\Delta S^\ddagger$ J/mol K	Ref.
HO$^-$	0.22, 0.55	52.7	− 75	[116,159]
CH$_3$CH$_2$NH$_2$	1.95 × 10^{-2}			[119]
HS$^-$	1.7 × 10^2	30.1	− 100	[123]
[SO$_3$]$^{2-}$	4.5 × 10^2	16.9	− 113	[160]
N$_3^-$	0.24			[161]
CH$_3$(CO$^-$)=CH$_2$	4.6 × 10^5			[126]
HO(CH$_2$)$_2$S$^-$	2.6 × 10^4	31.4	− 54	[53]
OAcNHC(CO$_2^-$)HCH$_2$S$^-$	3.4 × 10^4	35.2	− 44	[53]
$^-$O$_2$CCH$_2$C(CO$_2^-$)HS$^-$	4.8 × 10^3	33.9	− 59	[53]

kJ/mol, $\Delta S° = 105.5$ J/mol K, rate $= -1.4 \times 10^{-4}$ ($[Fe(CN)_5NO_2^{4-}]$ $[H_2O]$) [116]. Hydroxide similarly reacts with $[(NO)(CN)_5Ru]^{2-}$ ($K = 0.95$ M^{-1} S^{-1} $\Delta H^{\ddagger} = 57.3$ kJ/mol, $\Delta S^{\ddagger} = -54$ J, $K = 4.4 \times 10^6$); however, both nitro and nitrito linkage isomers are formed and the dissociation of nitrite to yield $[(H_2O)(CN)_5Ru]^{3-}$ is more rapid ($k = 2 \times 10^{-4}$ s^{-1}) [118].

Hydroxide attack on complexes of the type cis-$[(NO)L(bpy)_2Ru^{II}]$ also yields the corresponding nitro complex [118]. The equilibrium constants for these reactions are a strong function of the cis-ligand and increased by 10^{11} on changing L from a π-donor (Cl^-) to a π-acceptor (py) [98]. The ability of the nitro complexes to transfer an oxygen atom to a reductant such as triphenyl-phosphine is also a function of the cis-ligand [98].

Primary and secondary amines attack the nitrosyl nitrogen via pathways that are both first and second order in the amine concentration [26, 119]. This is followed by dissociation to give primary alcohols or nitrosoamines [120, 121]; however, tertiary amines and sterically hindered amines do not react.

$$[NO(CN)_5Fe]^{2-} + 2RNH_2 \rightarrow [RNH_2(CN)_5Fe]^{3-} + ROH + N_2$$

$$[NO(CN)_5Fe]^{2-} + 2R_2NH \rightarrow [R_2NH(CN)_5Fe]^{2-}$$
$$+ R_2NNO + H^+$$

The adduct formed, $[(CN)_5FeNONRR']^{2-}$, dissociates to yield a cationic nit-rosamine through either a dissociative or an interchange mechanism, with rate constants on the order of 10^{-3} M^{-1} s^{-1} and 10^{-3} M^{-2} s^{-1}, respectively [7, 119, 122]. Amino acids undergo similar nitrosation reactions to yield hydroxy acids or lactones. Diamino acids tend to yield cyclic imino acids, so that ornithine and lysine give rise to proline and pipecolic acid, respectively.

Thiolates react with nitroprusside to form the corresponding thionitro complex, which is often a precursor to electron transfer to yield $[Fe(CN)_5(NO)]^{3-}$ and the corresponding disulfide. Thiols with moderate to strong electron withdrawing groups (thioglucose, thionitrocresol, thiocyanate, thiourea), which significantly decrease the thiolate's nucleophilicity, often do not react with nitroprusside [53]. Protonation of the cysteine amine has little affect on the rate of adduct formation; however, it significantly increases the first-order rate constant for dissociation of the adduct back to starting materials from 0.03 s^{-1} to 0.18 s^{-1}. At pH 6.5–8.5, dissociation of thionitroso derivatives, RSNO, also occurs, which then decay homolytically to NO and the organic disulfide [54]. While the formation of nitrosothiols might occur in vivo to fix and then release NO, this is not a likely mechanism, since both adduct formation and decomposition are slower than reduction of nitroprusside and subsequent release of NO.

The rate of adduct formation with thiolate (A, below) is first order in nitroprusside and thiolate (see Table 4), while the bimolecular rate constant for the second reaction is 1.3×10^{-2} M^{-1} s^{-1}, $\Delta H° = 81.2$ kJ/mol, $\Delta S° = 12.5$ J/mol K) [123]. As the rate of the second step ($t_{1/2} = 55$ s) is unusually slow for a proton transfer and is independent of pH between 11.5 and 12.8, it may be that a rate determining N to S linkage isomerization of the NOS^- occurs [3, 123].

1) $[Fe(CN)_5NO]^{2-} + SH^- \longrightarrow \left[Fe(CN)_5N \overset{\displaystyle O}{\underset{\displaystyle SH}{\diagdown}} \right]^{3-}$ (A)

2) $\left[Fe(CN)_5N \overset{\displaystyle O}{\underset{\displaystyle SH}{\diagdown}} \right]^{3-} + OH^- \longrightarrow \left[Fe(CN)_5N \overset{\displaystyle O}{\underset{\displaystyle S^-}{\diagdown}} \right]^{4-} + H_2O$

(B)

Since thiolates are far more reactive than thiols (pK_a 8-9.5) in nucleophilic attack on coordinated NO, these rates are pH dependent [53]. The reaction of alkaline solutions of mercaptans with nitroprusside to yield reddish-purple solutions has long been used as a test for cysteine, glutathione or other thiol-containing compounds. However, if the solution is too basic, the nitroprusside/hydroxide reaction becomes competitive [3].

$$[Fe(CN)_5NO]^{2-} + RS^- \rightarrow [Fe(CN)_5NO(SR)]^{3-}$$

A deep red coloration occurs in the Boedeker reaction when solutions of SO_3^{2-} and nitroprusside are combined. The color gradually fades as $[(CN)_5Fe(SO_3)]^{5-}$ is formed. The equilibrium is [3]:

$$k_f = 107 \, M^{-1} s^{-1}$$

$$[(CN)_5Fe(NO)]^{2-} + SO_3^{2-} \rightleftarrows [(CN)_5FeNO(SO_3)]^{4-}$$

$$k_r = 820 \, s^{-1}$$

Thiopental, which is commonly used as an anesthetic simultaneous with nitroprusside infusion, reacts as a thione with photoactivated nitroprusside to form an adduct analogous to that of thiourea, $[Fe^{II}(CN)_5(S = C-R)]^{3-}$, which is then oxidized to the ferric form, in a reaction which negates the desired effects of both the anesthetic and the hypotensive agent [7, 124, 125].

Slightly alkaline solutions containing ketones or other "acidic" hydrocarbons often react with nitroprusside to give a red coloration, which rapidly fades. The resulting solution contains $[(CN)_5Fe(H_2O)]^{3-}$ and the oxime of the organic compound [126].

1) $CH_3 \overset{\displaystyle O}{\overset{\displaystyle \|}{C}} CH_3 + OH^- \rightleftharpoons CH_3 \overset{\displaystyle O}{\overset{\displaystyle \|}{C}} CH_2^- + H_2O$

2) $[Fe(CN)_5NO]^{2-} + CH_3 \overset{\displaystyle O}{\overset{\displaystyle \|}{C}} CH_2^- \longrightarrow [(CN)_5FeNO{=}CH \overset{\displaystyle O}{\overset{\displaystyle \|}{C}} CH_3]^{4-} + H^+$

3) $(CN)_5FeNO[{=}CH \overset{\displaystyle O}{\overset{\displaystyle \|}{C}} CH_3]^{4-} \rightleftharpoons [(CN)_5Fe(H_2O)]^{3-} + CH_3 \overset{\displaystyle O}{\overset{\displaystyle \|}{C}} C{=}NOH$
                                                                                                                   $\underset{\displaystyle H}{|}$

The intermediate acetone adduct is reddish in color and forms with a bimolecular rate constant of 5×10^5 M^{-1} s^{-1}. This intermediate decomposes according to the second order rate law: rate $= 2.9 \times 10^{-4}$ [Fe] [H_2O] at 298 K with $\Delta H° = 74.5$ kJ/mol, $\Delta S° = -62.8$ J/mol K) [3, 126–129]. When secondary amines act as the base, the reaction proceeds through an enamine intermediate to yield complexes of the type: $[Fe(CN)_5NOCHRCOR']^{3-}$ [130]. At physiological pH, these reactions are extremely slow.

Nitroprusside forms both dinuclear and trinuclear complexes with aquacobalamin (B_{12a}) [131]. Coordination involves a Fe-C=N-Co link, with linkage involving trans cyanides in the trinuclear complex. Such complexes are slower acting as hypotensive agents than nitroprusside itself [7].

5.8 Nitroprusside Oxidation of Protein Sulfhydryls

The ability of coordinated NO to react with thiols has led to the suggestion of an alternative mechanism for activating guanylate cyclase. This involves nitroprusside oxidation of protein sulfhydryls to cross-link the protein with a disulfide bridge. For example, papain, which has an essential cysteine (cys-25) and glyceradehyde-3-phosphate dehydrogenase (cys-149) are both inhibited by nitroprusside with formation of $[Fe(CN)_5(NO)]^{3-}$ and $[Fe(CN)_4NO]^{2-}$ [132]. The suggested anaerobic reaction is:

$$Fe(CN)_5NO^{2-} + RS^- \rightleftharpoons \left[Fe(CN)_5 N \begin{matrix} O \\ SR \end{matrix} \right]^{3-} \longrightarrow [Fe(CN)_5 NO]^{3-} + RS\cdot$$

$$[Fe(CN)_6]^{4-} + Fe^{2+} + NO \quad RS–SR$$

Since hemoproteins such as lactoperoxidase and catalase are inhibited more rapidly than the sulfhydryl oxidation occurs, it is unlikely that the rapid activation of guanylate cyclase occurs by sulfhydryl oxidation [132]. Prolonged incubation of the papain or dehydrogenase enzymes with substrate and nitroprusside yielded a turbidity which indicated denaturation of the enzyme to an insoluble form, possibly by the formation of disulfide bridges via the dimerization of thiyl radicals [132].

6 Enzymatic Interactions of Nitric Oxide

6.1 Reactions of NO with Hemes and Guanylate Cyclase

Guanylate cyclase catalyzes the conversion of GTP to cGMP and exists in two forms: 1) a particulate form, which is membrane bound, and 2) a soluble form, which is free in the cytoplasm. The isozymes are distinct in that they: 1) exhibit different immunological reactivity, 2) derive from different genes, and 3) have different modes of regulation. Only the soluble enzyme contains a heme and is regulated by nitric oxide. However, arachidonic acid also activates the soluble form, particularly when it is heme-deficient [65]. Soluble guanylate cyclase as isolated from rat or bovine lung is a heterodimeric protein consisting of 70 kDa and 80 kDa subunits and contains a single Fe^{II} protoporphyrin IX moiety [133, 134]. It has a high affinity for protoporphyrin IX (K_d = 1.4 mM) through interactions involving both the hydrophobic side chains and the vicinal propionate groups. Heme is less tightly bound and can be reversibly dissociated by relatively mild methods [65]. The heme-deficient enzyme is not activated by the addition of heme, but protoporphyrin IX above stimulates activity about 50-fold over the basal rate [51].

Nitric oxide exerts a regulatory effect on the enzyme by interacting at the heme thereby generating a substantial increase in V_{max} for cGMP production and a decrease in the K_m for GTP [50]. Partial nitrification of guanylate cyclase appears to result in loss of the heme with concomitant loss of NO activation [135]. Since direct nitrosyl-heme transfer from other nitrosyl hemoproteins occurs, guanylate cyclase has a particularly high affinity for nitrosyl-heme relative to other hemoproteins. Activity is enhanced by ~50-fold when nitrosyl heme is added to the apoenzyme and ~60-fold when either NO or NO-heme is available to the holoenzyme [51]. The enzyme is not activated by nitroprusside alone, but activation ensues when a reducing thiol is added [65]. While it remains to be determined whether in vivo activation of guanylate cyclase in the cell is through: 1) solvated NO, 2) thionitroso compounds, or 3) nitrosylhemoproteins, at least in the brain solvated NO seems the most reasonable as direct heme coordination by this small, diffusible molecule should proceed much more rapidly than reactions involving intermediate formation with other hemoproteins or sulfur carriers.

Comparison of the NO-heme structure with that of nitroprusside (cf. Fig. 7 and 2) reveals an essential difference in structures involving complexes formulated as $Fe^{II}-NO^+$ and $Fe^{II}-NO\cdot$, in which the Fe–N–O bond angle of the former is considerably greater than that of the latter. Remarkably, EPR evidence indicates that this bond angle decreases to 110° at 77 K in nitrosyl hemes [138]. Theoretical calculations suggest two ground states close in energy, either of which may be influenced by protein interactions. At low temperature, the unpaired electron resides in mainly the iron d_{z^2} orbital ($Fe^{III}-NO^-$), while at room temperature, the odd electron is localized in an NO π^* orbital ($Fe^{II}-NO\cdot$) [139]. An analogous electronic situation may explain the unusual linear Fe–NO core in $[NO(CN)_4Fe]^{2-}$ (Fig. 3).

Coordination of NO to a heme involves appreciable π-backbonding between the Fe^{II} and the NO$\cdot$ such that a strong trans influence is exerted and the Fe is pulled toward the NO. This is evident in Fig. 7, where (Fig. 7a) the Fe–N(imidazole) is long compared to the usual value of ~1.98 Å and (Fig. 7b) where the imidazole is lost completely. Comparison of the two structures in

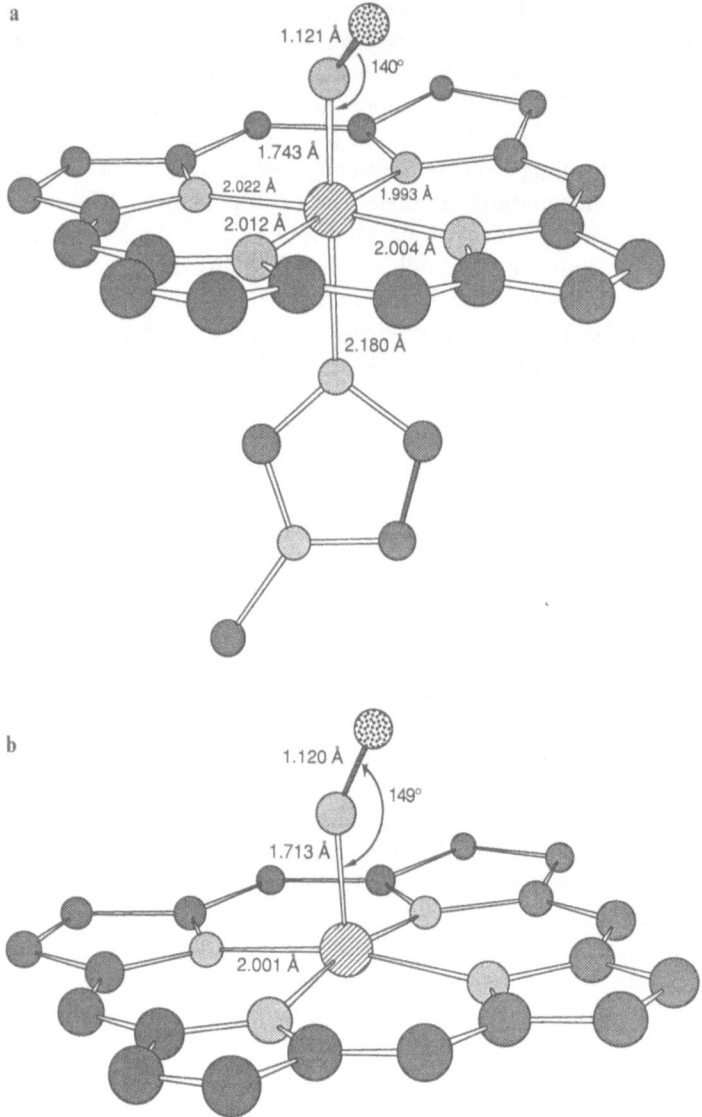

Fig. 7a, b. Structural changes between **a)** NO-Fe[II]-TPP-1MeIm [136] and **b)** NO-Fe[II]-TPP [137], where TPP = tetraphenylporphyrin (phenyl rings omitted)

Fig. 7 also shows that base binding in the *trans* position weakens the metal to nitrosyl backbonding in that the Fe–NO bond length increases by 0.3 Å and the Fe–N–O bond angle decreases ∼9°. In concert with this, the displacement of the Fe from the porphyrin plane decreases from 0.21 Å in the square pyramidal structure to 0.07 Å in the octahedral complex. The equilibria of NO binding to hemes is summarized in Fig. 8 [64], which indicates that NO binds to Fe[II] better

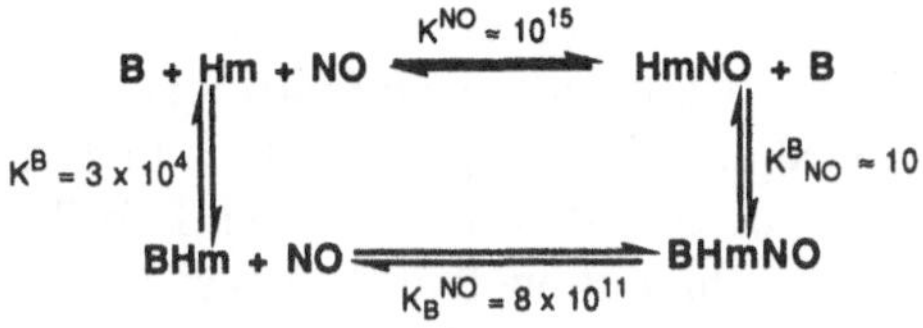

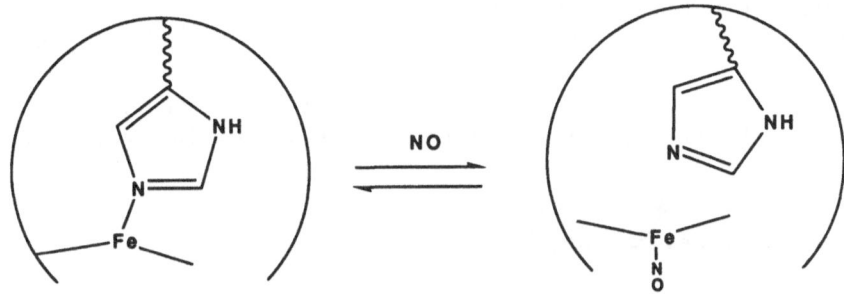

$$B + Hm + NO \xrightleftharpoons[]{K^{NO} \approx 10^{15}} HmNO + B$$

$$K^B = 3 \times 10^4 \updownarrow \qquad \qquad \updownarrow K^B_{NO} \approx 10$$

$$BHm + NO \xrightleftharpoons[K_B^{NO} = 8 \times 10^{11}]{} BHmNO$$

Fig. 8. Equilibria between heme, NO and imidazole

Fig. 9. Schematic of effect of nitrosyl coordination on a heme in a hemoprotein [64]

when there is no heterocyclic base, such as imidazole, in the *trans* position. The *trans* influence of the NO is such that the binding constant for the imidazole base (B) is decreased by a factor of $\sim 10^3$. This provides a driving force for the release of the proximal imidazole in hemoglobin upon coordination of NO [64, 139].

Traylor [64] and Ignarro [65] have suggested that the release of imidazole from the heme in guanylate cyclase could either loosen the protein for a conformational change or leave the histidine group available for catalytic service (see Fig. 9). The nitrosylheme might also move to a new site within the protein. At neutral pH and below, an additional driving force provided by protonation of the imidazole would facilitate heme release. At pH 4, nitrosyl-myoglobin completely loses its proximal base to free the protoheme-NO within its protein pocket. Some or all of these three effects could play key roles in the activation of guanylate cyclase. NO-hemes are much more stable in a kinetic sense than the isoelectronic carbonyl complexes, as the first-order dissociation rate constant for heme-CO is $10^4\,s^{-1}$ while that for NO-heme is $10^{-6}\,s^{-1}$. This affords the NO-heme sufficient time as an entity to move within the protein or diffuse freely through the solution. In contrast, CO-heme not only persists for a shorter period, but has a high affinity for the proximal base and so remains tethered to the protein [64].

6.2 Nitric Oxide Synthase

NO-synthase (EC 1.14.23) catalyzes the oxidation of arginine by O_2 to citrulline through a five-electron oxidation process, which gives the appearance of an intermolecular dioxygenase activity (Fig. 10). Complementary DNAs for both brain [10], macrophage [140] and endothelial [141] NO synthase have been cloned and there appears to be a high degree of homology and function in all forms investigated to date [12]. Labelling studies with $^{18}O_2$ and macrophage NO-synthase show that the heavier isotope appears in both oxidized products, NO and citrulline [11, 43, 66, 72]. Brain NO-synthase is a 150 kDa, heme-centered flavoprotein, which appears to have some P450-like characteristics [10, 142]. It has a recognition site for arginine, and is also regulated by interactions with several other molecules [143]. The enzyme has an absolute requirement for calmodulin, requires NADPH and has tightly bound flavins (FAD and FMN). A tetrahydropterin, 6(R)-tetrahydro-L-biopterin, is a required cofactor for both macrophage and cerebellar NO-synthase [66, 144]. The brain

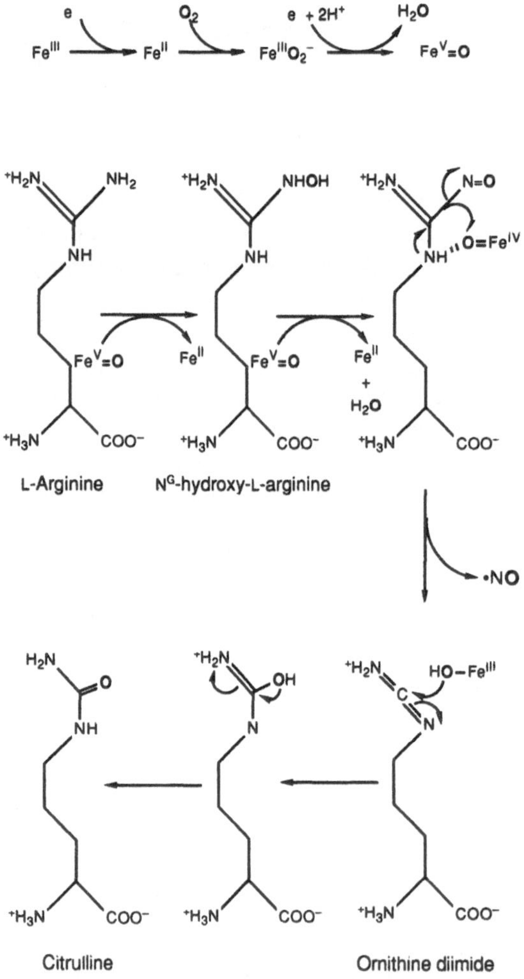

Fig. 10. Possible series of reactions for the biosynthesis of NO and citrulline from arginine

enzyme is stoichiometrically phosphorylated by cAMP-dependent protein kinase, protein kinase C and Ca^{2+}-calmodulin protein kinase [10, 61]. As with P450 enzymes, the requirement for a reductant (NADPH) may be necessary for activating the enzyme by reducing the iron so that O_2 can be coordinated as $Fe^{III}-O_2^-$ and then as $Fe^{III}-O_2^{2-}$ prior to oxygen atom transfer. Precedent for heme-centered dioxygenase activity occurs in indoleamine 2,3-dioxygenase, whose active site, like P450, involves a protoheme [145, 146]. The bound flavins, which often serve as electron-sinks and interfaces between single and double electron transfer, probably facilitate unit adjustments in the iron oxidation state at appropriate stages in the enzymatic cycle (see Fig. 10) from the two-electron donor, NADPH.

As the brain NO-synthase is highly homologous with cytochrome P450 reductase, it appears that the enzyme carries out its own reductase function to activate oxygen, and the electron transfer sequence might be expected to similar in the two enzymes [142, 147].

$$NADPH \rightarrow [FAD \rightarrow FMN] \rightarrow heme$$

This electron transfer pathway may partially explain the extraordinarily high efficacy of nitroprusside as it provides single electrons at a sufficiently negative potential to reduce both P450 ($E° = -0.17$ to -0.35 V) [148] and nitroprusside. Since NADPH/P450 reductase has been shown to be an efficient activator of this drug [43], NO may be evolved by NO-synthase reduction of nitroprusside at the same sites where NO is produced physiologically.

Given the analogies with P450, it is likely that NO-synthase functions as a monooxygenase to hydroxylate the arginine nitrogen as shown in Fig. 10. However, unlike the monooxygenase enzymes, oxygen from O_2 is also incorporated into two components of the organic substrate to yield citrulline and NO. This may involve sequential oxygen atom transfers (as shown) or a second site, such as a pterin, to allow an oxygen from O_2 to attack the organic substrate. Little is known about the mechanism following the first hydroxylation to form N^G-hydroxy-L-arginine, but a one-electron oxidation of the substrate coupled in some way with a single electron flavin reduction of the $[Fe^V = O]^+$ core is one possibility. It may be significant that at least one form of epilepsy, a disease affected by nitric oxide production and involving NMDA receptors [149], maps to the same area of chromosome 12 in rats that regulates P450 production [150].

6.3 Nitric Oxide Production by Denitrifying Bacteria

Nitric oxide and iron nitrosyl complexes have been observed in the reduction of nitrite by bacterial nitrite reductases, which contain iron chlorin or iron isobacterichlorin [151]. A specific nitric oxide reductase also exists to convert NO to nitrous oxide [9]. Iron complexes of chlorins, isobacteriochlorins, and porphyrins, as well as ruthenium and osmium polypyridines, and cobalt and nickel cyclams also reduce nitrite with concommitant formation of nitrosyl complexes [22]. Iron(III) porphyrins can coordinate NO to form iron(II)-nitrosyls, which are easily reduced to Fe^{II}–NO· complexes around 0.47 V. Nitrite can be electrocatalytically reduced through a nitrosyl intermediate to ammonia in the presence of Fe^{III}(TMPyP), where TMPyP = tetrakis(N-methyl-2-pyridyl)porphine, at a relatively modest potential (− 0.38 V) [22]. Oxidation of Fe^{II}–NO complexes of octaethylchlorin (OEC) and octaethylisobacteriochlorin (OEiBC) results in π-cation radials of these complexes, which can be induced to undergo valence isomerization from $(OEC^+)Fe^{II}$–NO to $(OEC)Fe^{II}$–NO^+ [151].

7 Bacteriostatic Effects

Preservation from *clostridia botulinum* colonization is effected by heat-curing meats with nitrite concentrations of ∼50 mg/kg (725 μM) [152, 153]. An unknown species, which is known as the "Perigo factor", is formed by heating nitrite in clostridial growth medium [154]. It has been suggested that this substance(s) involves iron-sulfide-nitrosyl complexes, since cluster compounds including $[Fe_4S_3(NO)_7]$ and $[Fe_4S_4(NO)_4]$ effectively inhibit clostridial growth [14]. Moreover, $Na[Fe_4S_3(NO)_7]$ and $[Fe_2(SR)_2(NO)_4]$ form on heating Fe^{II}, NO_2^- and biological sulfur (cystein, glutathione, peniccillamine, hydrolyzed casein, etc.) [115, 155]. It may be that such clusters: 1) interfere with the assembly of iron-sulfur clusters into proteins, 2) transfer nitrosyl to iron-sulfur clusters in proteins thereby inhibiting their effectiveness, or 3) serve as a source of nitric oxide to inhibit heme proteins [154]. Notably, the methyl ester of Roussin's red complex, μ-$(CH_3S)_2[Fe(NO)_2]_2$, nitrosates secondary ammines only in the presence of air, suggesting that it is air-oxidized to release NO [26]. The complexes $[Fe_2S_2(NO)_4]^{2-}$ and $[Fe_4S_3(NO)_7]^-$ undergo sequential, reversible one-electron reductions, but their oxidation couples are complex and irreversible [156]. Nitrosyl dithiocarbamato complexes of iron, $[Fe(NO)(S_2CNR_2)_2]$, undergo a reversible one-electron reduction to intermediates which readily lose NO [156].

8 Conclusion

The wide-spread and diverse functions of NO in biology can be mediated through metallonitrosyl complexes. This opens the way for the coordination chemist to design new agents as: 1) cytotoxic drugs that are activated by reduction in vivo to release NO; 2) more versatile hypotensive drugs; 3) facilitators of penile erection that might be applied topically, possibly in conjunction with prophylactics; 4) modulators of long-term potentiation at the initiation of memory in the hippocampus; and 5) facilitators of motor function coordination in the cerebellum. The factors involved in the development of these pharmaceuticals are well known to transition metal pharmaceutical chemists and involve ligand variations to modulate: 1) lipophilicity, 2) charge, 3) reduction potential, and 4) π-bonding and antibonding energy levels to weaken or strengthen the metallonitrosyl bond so as to facilitate the release or retention of NO. The first two parameters are generally more important in consideration involving cellular uptake, while the latter two will probably predominate in regulating intracellular NO release. Now that a burgeoning body of knowledge exists on the biological chemistry of NO and a horizon of pharamaceutical possibilities is in sight, rapid developments of this area can be expected.

Acknowledgement. This work was supported by NIH grant GM26390.

9 References

1. Ignarro LJ (1992) J NIH Resch 4: 59
2. Sundquist WI, Lippard SJ (1990) Coord Chem Rev 100: 293
3. Swinehart JH (1967) Coord Chem Rev 2: 385–402
4. Playfair L (1849) Proc Roy Soc (London)5: 846
5. Johnson CC (1929) Arch Int Pharmacodyn Ther 35: 489
6. Moraca PP, Bilte EM, Hale DE, Wasmuth CE, Pontasse EF (1962) Anesthesiology 23: 193
7. Butler AR, Glidewell C (1987) Chem Soc Rev 16: 361–380
8. Greenwood NN, Earnshaw A (1984) In: Chemistry of the elements. Pergamon, New York
9. Ye RW, Toro-Suarez I, Tiedji JM (1991) J Biol Chem 266: 12848
10. Bredt DS, Hwang PM, Glatt CE, Lowenstein C, Reed RR, Synder SH (1991) Nature 351: 714–718
11. Moncada S, Palmer RMJ, Higgs EA (1991) Pharmacological Reviews 43: 109–142
12. Snyder S (1992) Science 257: 494–496
13. Snyder SH, Bredt DS (1992) Scientific American 68–77
14. Payne MJ, Glidewell C (1990) J Gen Microbiol 136: 2077–87
15. Jones K (1973) In: Comprehensive inorganic chemistry. Pergamon, Oxford
16. Pedley JB, Marshal EM (1983) J Phys Chem Ref Data 12: 988
17. Beattie IR (1967) In: Mellor JW Comprehensive treatise on inorganic and theoretical chemistry. Wiley, New York
18. Neta P, Huie RE, Ross AB (1988) J Phys Chem Ref Data 17: 1112
19. Simon A (1992) Angew Chem 31: 301
20. Shibuki K (1990) Neuroscience Research 9: 69–76
21. Malinski T, Taha Z (1992) Nature 358: 676
22. Chen S-M, Su YO (1990) J Electroanalyt Chem Interfac Phenom 280: 189
23. Beckman JS, Beckman TW, Chen J, Marshall P, Freeman BA (1990) Proc Natl Acad Sci USA 87: 1620–1624
24. Korth H-G, Ingold KU, Sustman R (1992) Angew Chem 31: 891
25. Pryor WA, Chruch DF, Govindan CK, Crank G (1982) J Org Chem 47: 156
26. Williams DLH (1988) In: Nitrosation. Cambridge University Press, New York
27. Roche L (1974) Nipride Product Report, Nutley, NJ
28. Tuzel IH (1974) J Clinic Pharmacol 14: 494–503
29. Abbot Laboratories, North Chicago, IL (1990) Nitropress™ package insert
30. Hale M (1990) Am J Nurs 90: 61–2
31. Hayes C (1992) Dept Pediatric Cardiology, Columbia Presbyterian Hospital (Personal communication)

32. Grosmaire EK (1992) Heart & lung 21: 214
33. Ikram H, Low CJS, Crozier IG, Shirlaw T (1992) Am J Cardiol 69: 361–366
34. Stroud S, Dyer J (1991) Am J Nurs 91: 78
35. Watt GD, Christensen J, Izatt RM (1965) Inorg Chem 9: 239
36. Butler AR, Glidewell C, Waddon AE (1989) Polyhedron 8: 2627
37. Rindone JA, Sloane EP (1992) Annals of Pharmacotherapy 26: 515
38. Kruszyna H, Kruszyna R, Hu J, Smith RP (1980) J Toxicol Environ Health 6: 757–773
39. Furchgott RF, Zawadzki JV (1980) Nature 288: 373–376
40. Ignarro LJ, Ross G, Tillisch J (1991) West J Med 154: 51
41. Garthwaite J (1991) TINS 14: 60–67
42. Clapp LH, Gurney AM (1991) Pflung Arch 418: 462–470
43. Rao RDN, Elguindi S, O'Brien PJ (1990) Arch Biochem Biophys 286: 30–37
44. Kruszyna HK Robert; Smith, Roger P; Wilcox, Dean E. (1987) Toxi App Pharma 91: 429–438
45. Kruszyna H, Kruszyna R, Smith RP, Wilcox DE (1988) Toxicol Appl Pharmacol 94: 458–465
46. Smith RP, Louis CA, Kruszyna RT, Kruszyna H (1991) Fund Appl Toxic 17: 120–127
47. Myers PR, Minor RL, Guerra R, Banes JN, Harrison DG (1990) Nature 345: 161–163
48. Ignarro LJ (1981) Biochem Phys Acta 673: 394
49. Ignarro LJ, Lippton H, Edwards JC, Baricos WH, Hyman AL, Kadowitz PJ, Gruetter CA (1981) Jour Pharm Exp Ther 218: 739–749
50. Ignarro LJ, Degnan JN, Baricos WH, Kadowitz PJ, Wolin MS (1982) Biochem Biophys Acta 718: 49–59
51. Ignarro LJ, Adams JB, Horwitz PM, Wood KS (1986) J Biol Chem 261: 4997–5002
52. Rubanyi GM, Johns A, Wilcox D, Bates FN, Harrison D (1991) J Cardiovasc Pharmacol 17 (Suppl. 3) : S41–S45
53. Johnson MD, Wilkins RG (1984) Inorg Chem 23: 231–235
54. Oae S, Kim YH, Fukushima D, Shinhama K (1978) J Chem Soc, Perkin Trans 1: 913
55. Burnett AL, Lowenstein CJ, Bredt DS, Chang TSK, Snyder SH (1992) Science 257: 401–403
56. Kots AY, Skurat AV, Sergienko EA (1992) FEBS Lett 300:9
57. Southam E, Garthwaite J (1991) Neuroscience Letters 130: 107–111
58. Ross A, Bredt D, Snyder SH (1990) Trends in Neurosciences 13: 216–222
59. Bredt DS, Snyder SH (1992) Neuron 8:3
60. Nowak R (1992) J of NIH Research 4: 49–55
61. Silva AJ, Stevens CF, Tonegaws S, Wang Y (1992) Science 257: 201–206
62. Hirsch JC, Crepel F (1992) Synapse 10: 173
63. Bredt DS, Hwang PM, Snyder SH (1990) Nature 347: 768
64. Traylor TG; S Vijay S. (1992) Biochem 31: 2847–2849
65. Ignarro LJ (1989) Seminars in hematology 26: 63
66. Marletta MA, Yoon PS, Iyengar R, Leaf CD, Wishnok JS (1988) Biochemistry 27: 8706–8711
67. Lancaster JR, Langrehr JM, Bergonia HA (1992) J Biol Chem 267: 10994
68. Kosaka H, Watanabe M, Yoshihara H, Narada N (1992) Biochem Biophys Res Comm 184: 1119
69. Mordvintcev P, Mulsch A, Busse R (1991) Analyt Biochem 199: 142
70. Hibbs JB, Taintor RR, Varvin Z (1988) Biochem Biophys Res Commun 157: 87–94
71. Moncada S, Hibbs EA (1990) Nitric oxide from L-arginine: A bioregulatory system. Elsevier, Amsterdam
72. Leone AM, Palmer RMJ, Knowles RG, Francis PL, Ashton DS, Monacada S (1991) J Biol Chem 266: 23790–23795
73. McCall TB, Feelisch M, Palmer RMJ, Moncada S (1991) Br J Pharmacol 102: 234–238
74. Selkoe DJ (1992) Neuron 6: 487–498
75. Glidewell C, Johnson IL (1987) Inorg Chim Acta 132: 145–147
76. Schmidt J, Kühr H, Dorn WI, Köpf J (1974) Inorg Nucl Chem Lett 10: 55–61
77. Symons MCR, West DX, Wilkinson JG (1976) Inorg Chem 15: 1022
78. McCleverty JA (1979) Chem Rev 79: 53–76
79. Armor JN, Hoffman MZ (1975) Inorg Chem 14: 444
80. Callahan RW, Brown GM, Meyer TJ (1975) J Am Chem Soc 97: 894–895
81. Castellano EE; R B.E.; Piro, E.O.; Amalvy, J.I. (1989) Acta Cryst C45: 1207–1210
82. Bottomley F, White PS (1979) Acta Cryst B35: 2193–95
83. Navaza A, Chevrier G, Alzari PM, Aymonino L (1989) Acta Cryst C45: 839–841
84. Olabe JA, Gentil LA, Rigotti G, Navaza A (1984) Inorg Chem 23: 4297–4302
85. Kolthoff IM, Toren PE (1953) J Amer Chem Soc 75: 1197–1201

86. Masek J, Maslova E (1974) Coll Czech Chem Commun 39: 2141–2161
87. Bowden WL, Bonnar P, Brown DB, Geiger WE (1977) Inorg Chem 16: 41
88. Gaul JB, Clarke MJ (1992)
89. Cheney RP, Simic MG, Hoffman MZ, Taub IA, Asmus KD (1977) Inorg Chem 16: 2187–2192
90. Cheney RP, Pell SD, Hoffman MZ (1979) J Inorg Nucl Chem 41: 489
91. Morando PJ, Borghi EB, de Schteingart LM, Blesa MA (1981) J Chem Soc, Dalton Trans 435
92. Butler AR, Glidewell C, Johnson IL, McIntosh AS (1987) Inorg Chim Acta 138: 159
93. McCleverty JA (1977) Chem Rev 77: 53–76
94. Eisenberg R, Meyer CD (1975) Acc Chem Res 8: 26–34
95. Lever ABP (1990) Inorg Chem 29: 1271–1285
96. Lu J, Clarke MJ (1992) J Chem Soc Dalton Trans 1243–1248
97. Lever ABP (1992) In: Molecular electrochemistry of inorganic, bioinorganic and organometallic compounds, Proceedings NATO Advanced Research Workshop, Sintra, Portugal
98. Pipes DW, Meyer TJ (1984) Inorg Chem 23: 2466–72
99. Armor JN (1970) PhD Thesis, Stanford University
100. Cheney R (1976) PhD Thesis, Boston University
101. Seddon EA, Seddon KR (1984) The chemistry of ruthenium, Elsevier, New York
102. Arnold WP, Longnecker DE, Epstein RM (1984) Anesthesiology 61: 254–260
103. Shafer PR, Wilcox DE, Kruszyna H, Kruszyna R, Smith RP (1989) Toxic Appl Pharmac 99:1–10
104. Wilcox DE, Kruszyna H, Kruszyna R, Smith RP (1990) Chem Res Toxicol 3: 71–76
105. Ungermann CB, Caulton KG (1976) J Am Chem Soc 98: 3862–3868
106. Richter-Addo GB, Legzdins P (1988) Chem Rev 88: 991
107. Doyle MP, Pickering RA, Dykstra RL, Cook BR (1982) J Am Chem Soc 104: 3392–3397
108. Mu XH, Kadish KM (1990) Inorg Chem 29: 1031
109. Stochel G, van Eldik R, Stasicka Z (1986) Inorg Chem 25: 3663–66
110. Stochel G (1992) Coord Chem Rev 114: 269–295
111. Bisset WIK, Burdon MG, Butler AR, Glidewell C, Reglinski J (1981) Br J Anaseth 53: 1015–1018
112. Wolfe SK, Swinehart JH (1975) Inorg Chem 14: 1049
113. Güdel HU (1990) Chem Phys Lett 175: 262–265
114. Fieldler J, Masek J (1981) Inorg Chim Acta 81: 117
115. Butler AR, Glidewell C, Glidewell 3M (1990) Polyhedron 9: 2399
116. Swinehart JH, Rock PA (1966) Inorg Chem 5: 573
117. Turney TA, Wright GA (1959) Chem Revs 59: 497–513
118. Chevalier AA (1991) J Chem Soc Dalt Trans 1959
119. Butler AR, Glidewell C, Reglinski J, Waddon A (1984) J Chem Res 279 (S) , 2768–2783 (M)
120. Maltz H, Grant MA, Navoroli MC (1971) J Org Chem 36: 363
121. McGarvey GJ, Kimura M (1986) J Org Chem 51: 3913
122. Dozsa L, Kormow V, Beck MT (1984) Inorg Chim Acta 82: 69
123. Rock PA, Swinehart JH (1966) Inorg Chem 5: 1078
124. Jaksevac-Miksa M, Hankony V, Karas-Gasparec V (1980) Z Phys Chem (Leipzig) 261: 1041
125. Glidewell C, Musgrave VAJ (1989) Inorg Chim Acta 167: 253–256
126. Swinehart JH, Schmidt WG (1967) Inorg Chem 6: 232
127. Wolfe SK, Swinehart JH (1968) Inorg Chem 7: 1855
128. Iha NYM, Toma HE (1984) Inorg Chim Acta 81: 181
129. Butler AR, Glidewell C, Chaipanich V, McGinnis J (1986) J Chem Soc, Perkin Trans 2:7
130. Wiegrebe W, Violbig M (1982) Z Naturforsch 37b: 490
131. Butler AR, Glidewell C, McIntosh AS (1986) Inorg Chem 25: 970–3
132. Butler AR, Calsy AM, Johnson IL (1989) Polyhedron 9: 913–919
133. Waldman SA, Murad F (1987) Pharm Rev 39: 163–196
134. Nakane M, Arai K, Saheki S, Kuno T, Buechler W, Murad F, (1990) J Biol Chem 265: 16841–16845
135. Craven PA, DeRubertis FR (1978) J Biol Chem 253: 8433–8443
136. Scheidt WR, Piciulo PL (1976) J Am Chem Soc 98: 1913–1919
137. Scheidt WR, Frisse ME (1975) J Am Chem Soc 97: 17–21
138. Hori H, Ikeda-Saito M, Yonetani T (1981) J Biol Chem 256: 7849–7855
139. Waleh A, Ho N, Chantranupong L, Loew GH (1989) J Am Chem Soc 111: 2767
140. Lyons CR, Orloff GJ, Cunningham JM (1992) J Biol Chem 267: 6370
141. Janssens SP, Shimouchi A, Quertermous T (1992) J Biol Chem 267: 14519

142. White KA, Marletta MA (1992) Biochem 3: 6627–6631
143. Bredt DS, Ferris CD, Snyder SH (1992) J Biol Chem 267: 10976
144. Mayer B, John M, Bohme E (1990) FEBS Lett 277: 215
145. Hayaishi O, Takikawa O, Yoshida R (1990) Prog Inorg Chem 38: 75–95
146. Cady SG, Sono M (1991) Arch Biochem Biophys 291: 326
147. Narayanasami R, Otvos JD, Kasper CB (1992) Biochem 31: 4210
148. Ortiz de Montellano P In: (eds) (1986) Plenum Press, New York
149. Wieraszko A, Seyfried TN (1991) Trans Am Soc Neurochem 22: 172
150. Seyfried TN, Glaser GH, Yu RK, Palayoor ST (1986) Adv Neurol 44: 115–133
151. Ozawa S, Jujji H, Morishima I (1992) J Am Chem Soc 114: 1548
152. Bonnet R, Chandra S, Charalambides AA, Sales K, Scourides PA (1980) J Chem Soc, Perkins Trans 1: 1706
153. van Roon PS (1980) Antonie van Leeuwenhoek 46: 515–16
154. Perigo JA, Roberts TA (1968) J Food Technol 3: 91–94
155. Butler AR, Glidewell C, Glidewell SM (1990) Polyhedron 11: 591–6
156. Crayston JA, Glidewell C, Lambert RJ (1990) Polyhedron 9: 1741
157. Bratsch SG (1989) J Phys Chem Ref Data 18: 13
158. Benderskii VA, Krivenko AG, Ponomarev EA (1990) Sov Electrochem 26: 285–291
159. Masek J, Wendt H (1969) Inorg Chim Acta 3: 455
160. Andrade C, Swinehart JH (1972) Inorg Chem 11: 648–650
161. Wolfe SK, Andrade C, Swinehart JH (1974) Inorg Chem 13: 2567–72
162. Stamler JS, Jaraki O, Osborne JA, Simon DI, Keaney J, Vita J, Singel DJ, Valeri CR, Loscalzo J (1992) Proc Natl Acad Sci (USA) 89: 7674–7677
163. Stamler JS, Simon DI, Osborne JA, Mullins ME, Jaraki O, Michel T, Singel DJ, Loscalzo J (1992) Proc Natl Acad Sci (USA) 89: 444–448
164. Zembowicz A, Vane JR (1992) Proc Natl Acad Sci (USA) 89: 2051
165. Kirchner JJ, Sigurdson ST, Hopkins PB (1992) J Am Chem Soc 114: 4021–4027
166. Wink DA, Kasprzak KS, Maragos CM (1991) Science Science: 1001
167. Moschel RC, Keefer LK (1989) Tetr Lett 30: 1467
168. Taha Z, Kiechle F, Malinski T (1992) Biochem Biophys Res Comm 188: 734–739

Note added in proof

The relatively acidic thiol on serum albumin forms a fairly stable S-nitroso adduct with nitric oxide, which may serve to preserve and carry NO throughout the circulatory system [162, 163]. Bacterial toxins released in toxic-shock syndrome induce excessive NO-synthase activity in macrophages. The resulting arterial expansion may induce the cardiovascular collapse associated with toxic shock syndrome [164]. Nitrous acid reacts with DNA to form dG-N2-dG interstrand crosslinks at the sequence 5'CG [165]. NO can also deaminate cytidine [166] and deoxyguanosine [167] and so may function as a mutagen. The rate law for NO reacting with O_2 has been measured electrochemically as [168]:

$$\frac{d[NO]}{dt} = k[O_2], \text{ where } k = 7.25 \times 10^{-3}\,s^{-1}$$

i.e. the half-life of NO is independent of [NO] and is ~ 4 min at $[O_2] = 50\,\mu M$. Since $t_{1/2}$ is considerably shortened in the presence of cells, it may be that amines and thiols are responsible for scavenging NO biologically.

Author Index Volumes 1–81

Ahrland, S.: Factors Contributing to (b)-behavior in Acceptors, Vol. 1, pp. 207–220.

Ahrland, S.: Thermodynamics of Complex Formation between Hard and Soft Acceptors and Donors. Vol. 5, pp. 118–149.

Ahrland, S.: Thermodynamics of the Stepwise Formation of Metal-Ion Complexes in Aqueous Solution. Vol. 15, pp. 167–188.

Allen, G. C., Warren, K. D.: The Electronic Spectra of the Hexafluoro Complexes of the First Transition Series. Vol. 9, pp. 49–138.

Allen, G. C., Warren, K. D.: The Electronic Spectra of the Hexafluoro Complexes of the Second and Third Transition Series. Vol. 19, pp. 105–165.

Alonso, J. A., Balbás, L. C.: Simple Density Functional Theory of the Electronegativity and Other Related Properties of Atoms and Ions. Vol. 66, pp. 41–78.

Alonso, J. A., Balbas, L. C.: Hardness of Metallic Clusters. Vol. 80, pp. 229–258.

Andersson, L. A., Dawson, J. H.: EXAFS Spectroscopy of Heme-Containing Oxygenases and Peroxidases. Vol. 74, pp. 1–40.

Ardon, M., Bino, A.: A New Aspect of Hydrolysis of Metal Ions: The Hydrogen-Oxide Bridging Ligand ($H_3O_2^-$). Vol. 65, pp. 1–28.

Armstrong, F. A.: Probing Metalloproteins by Voltammetry. Vol. 72, pp. 137–221.

Augustynski, J.: Aspects of Photo-Electrochemical and Surface Behavior of Titanium(IV) Oxide. Vol. 69, pp. 1–61.

Averill, B. A.: Fe–S and Mo–Fe–S Clusters as Models for the Active Site of Nitrogenase. Vol. 53, pp. 57–101.

Babel, D.: Structural Chemistry of Octahedral Fluorocomplexes of the Transition Elements Vol 3, pp. 1–87.

Bacci, M.: The Role of Vibronic Coupling in the Interpretation of Spectroscopic and Structural Properties of Biomolecules. Vol. 55, pp. 67–99.

Baekelandt, B. G., Mortier, W. J., Schoonheydt, R. A.: The EEM Approach to Chemical Hardness in Molecules and Solids: Fundamentals and Applications. Vol. 80, pp. 187–228.

Baker, E. C., Halstead, G. W., Raymond, K. N.: The Structure and Bonding of 4f and 5f Series Organometallic Compounds. Vol. 25, pp. 21–66.

Balsenc, L. R.: Sulfur Interaction with Surfaces and Interfaces Studied by Auger Electron Spectrometry. Vol. 39, pp. 83–114.

Banci, L., Bencini, A., Benelli, C., Gatteschi, D., Zanchini, C.: Spectral-Structural Correlations in High-Spin Cobalt(II) Complexes. Vol. 52, pp. 37–86.

Banci, L., Bertini, I., Luchinat, C.: The ^1H NMR Parameters of Magnetically Coupled Dimers — The Fe_2S_2 Proteins as an Example. Vol. 72, pp. 113–136.

Bartolotti, L. J.: Absolute Electronegativities as Determined from Kohn-Sham Theory. Vol. 66, pp. 27–40.

Baughan, E. C.: Structural Radii, Electron-cloud Radii, Ionic Radii and Solvation. Vol. 15, pp. 53–71.

Bayer, E., Schretzmann, P.: Reversible Oxygenierung von Metallkomplexen. Vol. 2, pp. 181–250.

Bearden, A. J., Dunham, W. R.: Iron Electronic Configuration in Proteins: Studies by Mössbauer Spectroscopy. Vol. 8, pp. 1–52.

Bergmann, D., Hinze, J.: Electronegativity and Charge Distribution. Vol. 66, pp. 145–190.

Berners-Price, S. J., Sadler, P. J.: Phosphines and Metal Phosphine Complexes: Relationship of Chemistry to Anticancer and Other Biological Activity. Vol. 70, pp. 27–102.

Bertini, I., Luchinat, C., Scozzafava, A.: Carbonic Anhydrase: An Insight into the Zinc Binding Site and into the Active Cavity Through Metal Substitution. Vol. 48, pp. 45–91.

Bertrand, P.: Application of Electron Transfer Theories to Biological Systems. Vol. 75, pp. 1–48.

Blasse, G.: The Influence of Charge-Transfer and Rydberg States on the Luminescence Properties of Lanthanides and Actinides. Vol. 26, pp. 43–79.

Blasse, G.: The Luminescence of Closed-Shell Transition Metal-Complexes. New Developments. Vol. 42, pp. 1–41.

Blasse, G.: Optical Electron Transfer Between Metal Ions and its Consequences. Vol. 76, pp. 153–188.

Blauer, G.: Optical Activity of Conjugated Proteins. Vol. 18, pp. 69–129.

Bleijenberg, K. C.: Luminescence Properties of Uranate Centres in Solids. Vol. 42, pp. 97–128.

Bŏca, R., Breza, M., Pelikán, P.: Vibronic Interactions in the Stereochemistry of Metal Complexes. Vol. 71, pp. 57–97.

Boeyens, J. C. A.: Molecular Mechanics and the Structure Hypothesis. Vol. 63, pp. 65–101.

Bonnelle, C.: Band and Localized States in Metallic Thorium, Uranium and Plutonium, and in Some Compounds, Studied by X-ray Spectroscopy. Vol. 31, pp. 23–48.

Bradshaw, A. M., Cederbaum, L. S., Domcke, W.: Ultraviolet Photoelectron Spectroscopy of Gases Adsorbed on Metal Surfaces. Vol. 24, pp. 133–170.

Braterman, P. S.: Spectra and Bonding in Metal Carbonyls. Part A: Bonding. Vol. 10, pp. 57–86.

Braterman, P. S.: Spectra and Bonding in Metal Carbonyls. Part B: Spectra and Their Interpretation. Vol. 26, pp. 1–42.

Bray, R. C., Swann, J. C.: Molybdenum-Containing Enzymes. Vol. 11, pp. 107–144.

Brese, N. E., O'Keeffe, M.: Crystal Chemistry of Inorganic Nitrides. Vol. 79, pp. 307–378.

Brooks, M. S. S.: The Theory of 5f Bonding in Actinide Solids. Vol. 59/60, pp. 263–293.

van Bronswyk, W.: The Application of Nuclear Quadrupole Resonance Spectroscopy to the Study of Transition Metal Compounds. Vol. 7, pp. 87–113.

Buchanan, B. B.: The Chemistry and Function of Ferredoxin. Vol. 1, pp. 109–148.

Buchler, J. W., Kokisch, W., Smith, P. D.: Cis, Trans, and Metal Effects in Transition Metal Porphyrins. Vol. 34, pp. 79–134.

Bulman, R. A.: Chemistry of Plutonium and the Transuranics in the Biospere. Vol. 34, pp. 39–77.

Bulman, R. A.: The Chemistry of Chelating Agents in Medical Sciences. Vol. 67, pp. 91–141.

Burdett, J. K.: The Shapes of Main-Group Molecules; A Simple Semi-Quantitative Molecular Orbital Approach. Vol. 31, pp. 67–105.

Burdett, J. K.: Some Structural Problems Examined Using the Method of Moments. Vol. 65, pp. 29–90.

Campagna, M., Wertheim, G. K., Bucher, E.: Spectroscopy of Homogeneous Mixed Valence Rare Earth Compounds. Vol. 30, pp. 99–140.

Ceulemans, A., Vanquickenborne, L. G.: The Epikernel Principle. Vol. 71, pp. 125–159.

Chandrasekhar, V., Thomas, K. R. Justin: Recent Aspects of the Structure and Reactivity of Cyclophosphazenes. Vol. 81, pp. 41–114.

Chasteen, N. D.: The Biochemistry of Vanadium, Vol. 53, pp. 103–136.

Chattaraj, P. K., Parr, R. G.: Density Functional Theory of Chemical Hardness. Vol. 80, pp. 11–26.

Cheh, A. M., Neilands, J. P.: The γ-Aminolevulinate Dehydratases: Molecular and Environmental Properties. Vol. 29, pp. 123–169.

Ciampolini, M.: Spectra of 3d Five-Coordinate Complexes. Vol. 6, pp. 52–93.

Chimiak, A., Neilands, J. B.: Lysine Analogues of Siderophores. Vol. 58, pp. 89–96.

Clack, D. W., Warren, K. D.: Metal-Ligand Bonding in 3d Sandwich Complexes. Vol. 39, pp. 1–141.

Clarke, M. J., Gaul, J. B.: Chemistry Relevant to the Biological Effects of Nitric Oxide and Metallonitrosyls. Vol. 81, pp. 147–181.

Clark, R. J. H., Stewart, B.: The Resonance Raman Effect. Review of the Theory and of Applications in Inorganic Chemistry. Vol. 36, pp. 1–80.

Clarke, M. J., Fackler, P. H.: The Chemistry of Technetium: Toward Improved Diagnostic Agents. Vol. 50, pp. 57–58.

Cohen, I. A.: Metal–Metal Interactions in Metalloporphyrins, Metalloproteins and Metalloenzymes. Vol. 40, pp. 1–37.

Connett, P. H., Wetterhahn, K. E.: Metabolism of the Carcinogen Chromate by Cellular Constituents. Vol. 54, pp. 93–124.

Cook, D. B.: The Approximate Calculation of Molecular Electronic Structures as a Theory of Valence. Vol. 35, pp. 37–86.

Cooper, S. R., Rawle, S. C.: Crown Thioether Chemistry. Vol. 72, pp. 1–72.
Cotton, F. A., Walton, R. A.: Metal–Metal Multiple Bonds in Dinuclear Clusters. Vol. 62, pp. 1–49.
Cox, P. A.: Fractional Parentage Methods for Ionisation of Open Shells of *d* and *f* Electrons. Vol. 24, pp. 59–81.
Crichton, R. R.: Ferritin. Vol. 17, pp. 67–134.

Daul, C., Schläpfer, C. W., von Zelewsky, A.: The Electronic Structure of Cobalt(II) Complexes with Schiff Bases and Related Ligands. Vol. 36, pp. 129–171.
Dehnicke, K., Shihada, A.-F.: Structural and Bonding Aspects in Phosphorus Chemistry-Inorganic Derivates of Oxohalogeno Phosphoric Acids. Vol. 28, pp. 51–82.
Denning, R. G.: Electronic Structure and Bonding in Actinyl Ions. Vol. 79, pp. 215–276.
Dobiáš, B.: Surfactant Adsorption on Minerals Related to Flotation. Vol. 56, pp. 91–147.
Doi, K., Antanaitis, B. C., Aisen, P.: The Binuclear Iron Centers of Uteroferrin and the Purple Acid Phosphatases. Vol. 70, pp. 1–26.
Doughty, M. J., Diehn, B.: Flavins as Photoreceptor Pigments for Behavioral Responses. Vol. 41, pp. 45–70.
Drago, R. S.: Quantitative Evaluation and Prediction of Donor-Acceptor Interactions. Vol. 15, pp. 73–139.
Drillon, M., Darriet, J.: Progress in Polymetallic Exchange-Coupled Systems, some Examples in Inorganic Chemistry. Vol. 79, pp. 55–100.
Dubhghaill, O. M. Ni, Sadler, P. J.: The Structure and Reactivity of Arsenic Compounds. Biological Activity and Drug Design. Vol. 78, pp. 129–190.
Duffy, J. A.: Optical Electronegativity and Nephelauxetic Effect in Oxide Systems. Vol. 32, pp. 147–166.
Dunn, M. F.: Mechanisms of Zinc Ion Catalysis in Small Molecules and Enzymes. Vol. 23, pp. 61–122.

Emsley, E.: The Composition, Structure and Hydrogen Bonding of the β-Diketones. Vol. 57, pp. 147–191.
Englman, R.: Vibrations in Interaction with Impurities. Vol. 43, pp. 113–158.
Epstein, I. R., Kustin, K.: Design of Inorganic Chemical Oscillators. Vol. 56, pp. 1–33.
Ermer, O.: Calculations of Molecular Properties Using Force Fields. Applications in Organic Chemistry. Vol. 27, pp. 161–211.
Ernst, R. D.: Structure and Bonding in Metal-Pentadienyl and Related Compounds. Vol. 57, pp. 1–53.
Erskine, R. W., Field, B. O.: Reversible Oxygenation. Vol. 28, pp. 1–50.
Evain, M., Brec, R.: A New Approach to Structural Description of Complex Polyhedra Containing Polychalcogenide Anions. Vol. 79, pp. 277–306.

Fajans, K.: Degrees of Polarity and Mutual Polarization of Ions in the Molecules of Alkali Fluorides, SrO, and BaO. Vol. 3, pp. 88–105.
Fee, J. A.: Copper Proteins – Systems Containing the "Blue" Copper Center. Vol. 23, pp. 1–60.
Feeney, R. E., Komatsu, S. K.: The Transferrins. Vol. 1, pp. 149–206.
Felsche, J.: The Crystal Chemistry of the Rare-Earth Silicates. Vol. 13, pp. 99–197.
Ferreira, R.: Paradoxical Violations of Koopmans' Theorem, with Special Reference to the 3d Transition Elements and the Lanthanides. Vol. 31, pp. 1–21.
Fidelis, I. K., Mioduski, T.: Double-Double Effect in the Inner Transition Elements. Vol. 47, pp. 27–51.
Fournier, J. M.: Magnetic Properties of Actinide Solids. Vol. 59/60, pp. 127–196.
Fournier, J. M., Manes, L.: Actinide Solids. 5f Dependence of Physical Properties. Vol. 59/60, pp. 1–56.
Fraga, S., Valdemoro, C.: Quantum Chemical Studies on the Submolecular Structure of the Nucleic Acids. Vol. 4, pp. 1–62.
Fraústo da Silva, J. J. R., Williams, R. J. P.: The Uptake of Elements by Biological Systems. Vol. 29, pp. 67–121.
Fricke, B.: Superheavy Elements. Vol. 21, pp. 89–144.

Fricke, J., Emmerling, A.: Aerogels–Preparation, Properties, Applications. Vol. 77, pp. 37–88.

Frenking, G., Cremer, D.: The Chemistry of the Noble Gas Elements Helium, Neon, and Argon – Experimental Facts and Theoretical Predictions, Vol. 73, pp. 17–96.

Fuhrhop, J.-H.: The Oxidation States and Reversible Redox Reactions of Metalloporphyrins. Vol. 18, pp. 1–67.

Furlani, C., Cauletti, C.: He(I) Photoelectron Spectra of *d*-metal Compounds. Vol. 35, pp. 119–169.

Gázquez, J. L., Vela, A., Galván, M.: Fukui Function, Electronegativity and Hardness in the Kohn-Sham Theory. Vol. 66, pp. 79–98.

Gazquéz, J. L.: Hardness and Softness in Density Functional Theory. Vol. 80, pp. 27–44.

Gerloch, M., Harding, J. H., Woolley, R. G.: The Context and Application of Ligand Field Theory. Vol. 46, pp. 1–46.

Gillard, R. D., Mitchell, P. R.: The Absolute Configuration of Transition Metal Complexes. Vol. 7, pp. 46–86.

Gleitzer, C., Goodenough, J. B.: Mixed-Valence Iron Oxides. Vol. 61, pp. 1–76.

Gliemann, G., Yersin, H.: Spectroscopic Properties of the Quasi One-Dimensional Tetra-cyanoplatinate(II) Compounds. Vol. 62, pp. 87–153.

Golovina, A. P., Zorov, N. B., Runov, V. K.: Chemical Luminescence Analysis of Inorganic Substances. Vol. 47, pp. 53–119.

Green, J. C.: Gas Phase Photoelectron Spectra of *d*- and *f*-Block Organometallic Compounds. Vol. 43, pp. 37–112.

Grenier, J. C., Pouchard, M., Hagenmuller, P.: Vacancy Ordering in Oxygen-Deficient Perovskite-Related Ferrites. Vol. 47, pp. 1–25.

Griffith, J. S.: On the General Theory of Magnetic Susceptibilities of Polynuclear Transitionmetal Compounds. Vol. 10, pp. 87–126.

Gubelmann, M. H., Williams, A. F.: The Structure and Reactivity of Dioxygen Complexes of the Transition Metals. Vol. 55, pp. 1–65.

Guilard, R., Lecomte, C., Kadish, K. M.: Synthesis, Electrochemistry, and Structural Properties of Porphyrins with Metal–Carbon Single Bonds and Metal–Metal Bonds. Vol. 64, pp. 205–268.

Gütlich, P.: Spin Crossover in Iron(II)-Complexes. Vol. 44, pp. 83–195.

Gutmann, V., Mayer, U.: Thermochemistry of the Chemical Bond. Vol. 10, pp. 127–151.

Gutmann, V., Mayer, U.: Redox Properties: Changes Effected by Coordination. Vol. 15, pp. 141–166.

Gutmann, V., Mayer, H.: Application of the Functional Approach to Bond Variations Under Pressure. Vol. 31, pp. 49–66.

Hall, D. I., Ling, J. H., Nyholm, R. S.: Metal Complexes of Chelating Olefin-Group V Ligands. Vol. 15, pp. 3–51.

Harnung, S. E., Schäffer, C. E.: Phase-fixed 3-Γ Symbols and Coupling Coefficients for the Point Groups. Vol. 12, pp. 201–255.

Harnung, S. E., Schäffer, C. E.: Real Irreducible Tensorial Sets and their Application to the Ligand-Field Theory. Vol. 12, pp. 257–295.

Hathaway, B. J.: The Evidence for "Out-of-the Plane" Bonding in Axial Complexes of the Copper(II) Ion. Vol. 14, pp. 49–67.

Hathaway, B. J.: A New Look at the Stereochemistry and Electronic Properties of Complexes of the Copper(II) Ion. Vol. 57, pp. 55–118.

Hellner, E. E.: The Frameworks (Bauverbände) of the Cubic Structure Types. Vol. 37, pp. 61–140.

von Herigonte, P.: Electron Correlation in the Seventies. Vol. 12, pp. 1–47.

Hemmerich, P., Michel, H., Schug, C., Massey, V.: Scope and Limitation of Single Electron Transfer in Biology. Vol. 48, pp. 93–124.

Henry, M., J. P. Jolivet, Livage, J.: Aqueous Chemistry of Metal Cations: Hydrolysis, Condensation and Complexation. Vol. 77, pp. 153–206.

Hider, R. C.: Siderophores Mediated Absorption of Iron. Vol. 58, pp. 25–88.

Hill, H. A. O., Röder, A., Williams, R. J. P.: The Chemical Nature and Reactivity of Cytochrome P-450. Vol. 8, pp. 123–151.

Hilpert, K.: Chemistry of Inorganic Vapors. Vol. 73, pp. 97–198.

Hogenkamp, H. P. C., Sando, G. N.: The Enzymatic Reduction of Ribonucleotides. Vol. 20, pp. 23–58.

Hoffman, B. M., Natan, M. J. Nocek, J. M., Wallin, S. A.: Long-Range Electron Transfer Within Metal-Substituted Protein Complexes. Vol. 75, pp. 85–108.

Hoffmann, D. K., Ruedenberg, K., Verkade, J. G.: Molecular Orbital Bonding Concepts in Polyatomic Molecules – A Novel Pictorial Approach. Vol. 33, pp. 57–96.

Hubert, S., Hussonnois, M., Guillaumont, R.: Measurement of Complexing Constants by Radiochemical Methods. Vol. 34, pp. 1–18.

Hudson, R. F.: Displacement Reactions and the Concept of Soft and Hard Acids and Bases. Vol. 1, pp. 221–223.

Hulliger, F.: Crystal Chemistry of Chalcogenides and Pnictides of the Transition Elements. Vol. 4, pp. 83–229.

Ibers, J. A., Pace, L. J., Martinsen, J., Hoffman, B. M.: Stacked Metal Complexes: Structures and Properties. Vol. 50, pp. 1–55.

Iqbal, Z.: Intra- und Inter-Molecular Bonding and Structure of Inorganic Pseudohalides with Triatomic Groupings. Vol. 10, pp. 25–55.

Izatt, R. M., Eatough, D. J., Christensen, J. J.: Thermodynamics of Cation-Macrocyclic Compound Interaction. Vol. 16, pp. 161–189.

Jain, V. K., Bohra, R., Mehrotra, R. C.: Structure and Bonding in Organic Derivatives of Antimony(V). Vol. 52, pp. 147–196.

Jerome-Lerutte, S.: Vibrational Spectra and Structural Properties of Complex Tetracyanides of Platinum, Palladium and Nickel. Vol. 10, pp. 153–166.

Jørgensen, C. K.: Electric Polarizability, Innocent Ligands and Spectroscopic Oxidation States. Vol. 1, pp. 234–248.

Jørgensen, C. K.: Heavy Elements Synthesized in Supernovae and Detected in Peculiar A-type Stars. Vol. 73, pp. 199–226.

Jørgensen, C. K.: Recent Progress in Ligand Field Theory. Vol. 1, pp. 3–31.

Jørgensen, C. K.: Relationship Between Softness, Covalent Bonding, Ionicity and Electric Polarizability. Vol. 3, pp. 106–115.

Jørgensen, C. K.: Valence-Shell Expansion Studied by Ultra-violet Spectroscopy. Vol. 6, pp. 94–115.

Jørgensen, C. K.: The Inner Mechanism of Rare Earths Elucidated by Photo-Electron Spectra. Vol. 13, pp. 199–253.

Jørgensen, C. K.: Partly Filled Shells Constituting Anti-bonding Orbitals with Higher Ionization Energy than Their Bonding Counterparts. Vol. 22, pp. 49–81.

Jørgensen, C. K.: Photo-electron Spectra of Non-metallic Solids and Consequences for Quantum Chemistry. Vol. 24, pp. 1–58.

Jørgensen, C. K.: Narrow Band Thermoluminescence (Candoluminescence) of Rare Earths in Auer Mantles. Vol. 25, pp. 1–20.

Jørgensen, C. K.: Deep-lying Valence Orbitals and Problems of Degeneracy and Intensities in Photo-electron Spectra. Vol. 30, pp. 141–192.

Jørgensen, C. K.: Predictable Quarkonium Chemistry. Vol. 34, pp. 19–38.

Jørgensen, C. K.: The Conditions for Total Symmetry Stabilizing Molecules, Atoms, Nuclei and Hadrons. Vol. 43, pp. 1–36.

Jørgensen, C. K., Frenking, G.: Historical, Spectroscopic and Chemical Comparison of Noble Gases. Vol. 73, pp. 1–16.

Jørgensen, C. K., Kauffmann, G. B.: Crookes and Marignac – A Centennial of an Intuitive and Pragmatic Appraisal of "Chemical Elements" and the Present Astrophysical Status of Nucleosynthesis and "Dark Matter". Vol. 73, pp. 227–254.

Jørgensen, C. K., Reisfeld, R.: Uranyl Photophysics. Vol. 50, pp. 121–171.

O'Keeffe, M.: The Prediction and Interpretation of Bond Lengths in Crystals. Vol. 71, pp. 161–190.

O'Keeffe, M., Hyde, B. G.: An Alternative Approach to Non-Molecular Crystal Structures with Emphasis on the Arrangements of Cations. Vol. 61, pp. 77–144.

Kahn, O.: Magnetism of the Heteropolymetallic Systems. Vol. 68, pp. 89–167.

Keppler, B. K., Friesen, C., Moritz, H. G., Vongerichten, H., Vogel, E.: Tumor-Inhibiting Bis (β-Diketonato) Metal Complexes. Budotitane, cis-Diethoxybis (1-phenylbutane-1,3-dionato) titanium (IV). Vol. 78, pp. 97–128.

Kimura, T.: Biochemical Aspects of Iron Sulfur Linkage in None-Heme Iron Protein, with Special Reference to "Adrenodoxin". Vol. 5, pp. 1–40.

Kitagawa, T., Ozaki, Y.: Infrared and Raman Spectra of Metalloporphyrins. Vol. 64, pp. 71–114.

Kiwi, J., Kalyanasundaram, K., Grätzel, M.: Visible Light Induced Cleavage of Water into Hydrogen and Oxygen in Colloidal and Microheterogeneous Systems. Vol. 49, pp. 37–125.

Kjekshus, A., Rakke, T.: Considerations on the Valence Concept. Vol. 19, pp. 45–83.

Kjekshus, A., Rakke, T.: Geometrical Considerations on the Marcasite Type Structure. Vol. 19, pp. 85–104.

König, E.: The Nephelauxelic Effect. Calculation and Accuracy of the Interelectronic Repulsion Parameters I. Cubic High-Spin d^2, d^3, d^7 and d^8 Systems. Vol. 9, pp. 175–212.

König, E.: Nature and Dynamics of the Spin-State Interconversion in Metal Complexes. Vol. 76, pp. 51–152.

Köpf-Maier, P., Köpf, H.: Transition and Main-Group Metal Cyclopentadienyl Complexes: Preclinical Studies on a Series of Antitumor Agents of Different Structural Type. Vol. 70, pp. 103–185.

Komorowski, L.: Hardness Indices for Free and Bonded Atoms. Vol. 80, pp. 45–70.

Koppikar, D. K., Sivapullaiah, P. V., Ramakrishnan, L., Soundararajan, S.: Complexes of the Lanthanides with Neutral Oxygen Donor Ligands. Vol. 34, pp. 135–213.

Krause, R.: Synthesis of Ruthenium(II) Complexes of Aromatic Chelating Heterocycles: Towards the Design of Luminescent Compounds. Vol. 67, pp. 1–52.

Krumholz, P.: Iron(II) Diimine and Related Complexes. Vol. 9, pp. 139–174.

Kuki, A.: Electronic Tunneling Paths in Proteins. Vol. 75, pp. 49–84.

Kustin, K., McLeod, G. C., Gilbert, T. R., Briggs, LeB. R., 4th.: Vanadium and Other Metal Ions in the Physiological Ecology of Marine Organisms. Vol. 53, pp. 137–158.

Labarre, J. F.: Conformational Analysis in Inorganic Chemistry: Semi-Empirical Quantum Calculation vs. Experiment. Vol. 35, pp. 1–35.

Lammers, M., Follmann, H.: The Ribonucleotide Reductases: A Unique Group of Metalloenzymes Essential for Cell Proliferation. Vol. 54, pp. 27–91.

Lehn, J.-M.: Design of Organic Complexing Agents. Strategies Towards Properties. Vol. 16, pp. 1–69.

Linarès, C., Louat, A., Blanchard, M.: Rare-Earth Oxygen Bonding in the LnMO₄ Xenotime Structure. Vol. 33, pp. 179–207.

Lindskog, S.: Cobalt(II) in Metalloenzymes. A Reporter of Structure-Function Relations. Vol. 8, pp. 153–196.

Liu, A., Neilands, J. B.: Mutational Analysis of Rhodotorulic Acid Synthesis in *Rhodotorula pilimanae*. Vol. 58, pp. 97–106.

Livorness, J., Smith, T.: The Role of Manganese in Photosynthesis. Vol. 48, pp. 1–44.

Llinás, M.: Metal-Polypeptide Interactions: The Conformational State of Iron Proteins. Vol. 17, pp. 135–220.

Lucken, E. A. C.: Valence-Shell Expansion Studied by Radio-Frequency Spectroscopy. Vol. 6, pp. 1–29.

Ludi, A., Güdel, H. U.: Structural Chemistry of Polynuclear Transition Metal Cyanides. Vol. 14, pp. 1–21.

Lutz, H. D.: Bonding and Structure of Water Molecules in Solid Hydrates. Correlation of Spectroscopic and Structural Data. Vol. 69, pp. 125.

Maggiora, G. M., Ingraham, L. L.: Chlorophyll Triplet States. Vol. 2, pp. 126–159.

Magyar, B.: Salzebullioskopie III. Vol. 14, pp. 111–140.

Makovicky, E., Hyde, B. G.: Non-Commensurate (Misfit) Layer Structures. Vol. 46, pp. 101–170.

Manes, L., Benedict, U.: Structural and Thermodynamic Properties of Actinide Solids and Their Relation to Bonding. Vol. 59/60, pp. 75–125.

Mann, S.: Mineralization in Biological Systems. Vol. 54, pp. 125–174.

March, N. H.: The Ground-State Energy of Atomic and Molecular Ions and Its Variation with the Number of Electrons. Vol. 80, pp. 71–86.

Mason, S. F.: The Ligand Polarization Model for the Spectra of Metal Complexes: The Dynamic Coupling Transition Probabilities. Vol. 39, pp. 43–81.

Mathey, F., Fischer, J., Nelson, J. H.: Complexing Modes of the Phosphole Moiety. Vol. 55, pp. 153–201.

Mauk, A. G.: Electron Transfer in Genetically Engineered Proteins. The Cytochrome *c* Paradigm. Vol. 75, pp. 131–158.

Mayer, U., Gutmann, V.: Phenomenological Approach to Cation-Solvent Interactions. Vol. 12, pp. 113–140.

Mazumdar, S., Mitra, S.: Biomimetic Chemistry of Hemes Inside Aqueous Micelles. Vol. 81, pp. 115–145.

McLendon, G.: Control of Biological Electron Transport via Molecular Recognition and Binding: The "Velcro" Model. Vol. 75, pp. 159–174.

Mehrotra, R. C.: Present Status and Future Potential of the Sol–Gel Process. Vol. 77, pp. 1–36.

Mildvan, A. S., Grisham, C. M.: The Role of Divalent Cations in the Mechanism of Enzyme Catalyzed Phosphoryl and Nucleotidyl. Vol. 20, pp. 1–21.

Mingos, D. M. P., Hawes, J. C.: Complementary Spherical Electron Density Model. Vol. 63, pp. 1–63.

Mingos, D. M. P., Johnston, R. L.: Theoretical Models of Cluster Bonding. Vol. 68, pp. 29–87.

Mingos, D. M. P., Zhenyang, L.: Non-Bonding Orbitals in Co-ordination Hydrocarbon and Cluster Compounds. Vol. 71, pp. 1–56.

Mingos, D. M. P., Zhenyang, L.: Hybridization Schemes for Co-ordination and Organometallic Compounds. Vol. 72, pp. 73–112.

Mingos, D. M. P., McGrady, J. E., Rohl, A. L.: Moments of Inertia in Cluster and Coordination Compounds. Vol. 79, pp. 1–54.

Moreau-Colin, M. L.: Electronic Spectra and Structural Properties of Complex Tetracyanides of Platinum, Palladium and Nickel. Vol. 10, pp. 167–190.

Morgan, B., Dophin, D.: Synthesis and Structure of Biomimetic Porphyrins. Vol. 64, pp. 115–204.

Morris, D. F. C.: Ionic Radii and Enthalpies of Hydration of Ions. Vol. 4, pp. 63–82.

Morris, D. F. C.: An Appendix to Structure and Bonding. Vol. 4 (1968). Vol. 6, pp. 157–159.

Mortensen, O. S.: A Noncommuting-Generator Approach to Molecular Symmetry. Vol. 68, pp. 1–28.

Mortier, J. W.: Electronegativity Equalization and its Applications. Vol. 66, pp. 125–143.

Müller, A., Baran, E. J., Carter, R. O.: Vibrational Spectra of Oxo-, Thio-, and Selenomellates of Transition Elements in the Solid State. Vol. 26, pp. 81–139.

Müller, A., Diemann, E., Jørgensen, C. K.: Electronic Spectra of Tetrahedral Oxo, Thio and Seleno Complexes Formed by Elements of the Beginning of the Transition Groups. Vol. 14, pp. 23–47.

Müller, U.: Strukturchemie der Azide. Vol. 14, pp. 141–172.

Müller, W., Spirlet, J.-C.: The Preparation of High Purity Actinide Metals and Compounds. Vol. 59/60, pp. 57–73.

Mullay, J. J.: Estimation of Atomic and Group Electronegativities. Vol. 66, pp. 1–25.

Murrell, J. N.: The Potential Energy Surfaces of Polyatomic Molecules. Vol. 32, pp. 93–146.

Naegele, J. R., Ghijsen, J.: Localization and Hybridization of 5f States in the Metallic and Ionic Bond as Investigated by Photoelectron Spectroscopy. Vol. 59/60, pp. 197–262.

Nag, K., Bose, S. N.: Chemistry of Tetra- and Pentavalent Chromium. Vol. 63, pp. 153–197.

Nalewajski, R. F.: The Hardness Based Molecular Charge Sensitivities and Their Use in the Theory of Chemical Reactivity. Vol. 80, pp. 115–186.

Neilands, J. B.: Naturally Occurring Non-porphyrin Iron Compounds. Vol. 1, pp. 59–108.

Neilands, J. B.: Evolution of Biological Iron Binding Centers. Vol. 11, pp. 145–170.

Neilands, J. B.: Methodology of Siderophores. Vol. 58, pp. 1–24.

Nieboer, E.: The Lanthanide Ions as Structural Probes in Biological and Model Systems. Vol. 22, pp. 1–47.

Novack, A.: Hydrogen Bonding in Solids. Correlation of Spectroscopic and Crystallographic Data. Vol. 18, pp. 177–216.
Nultsch, W., Häder, D.-P.: Light Perception and Sensory Transduction in Photosynthetic Prokaryotes. Vol. 41, pp. 111–139.

Odom, J. D.: Selenium Biochemistry. Chemical and Physical Studies. Vol. 54, pp. 1–26.
Oelkrug, D.: Absorption Spectra and Ligand Field Parameters of Tetragonal 3d-Transition Metal Fluorides. Vol. 9, pp. 1–26.
Oosterhuis, W. T.: The Electronic State of Iron in Some Natural Iron Compounds: Determination by Mössbauer and ESR Spectroscopy. Vol. 20, pp. 59–99.
Orchin, M., Bollinger, D. M.: Hydrogen-Deuterium Exchange in Aromatic Compounds. Vol. 23, pp. 167–193.

Peacock, R. D.: The Intensities of Lanthanide $f \leftrightarrow f$ Transitions. Vol. 22, pp. 83–122.
Pearson, R. G.: Chemical Hardness–An Historical Introduction. Vol. 80, pp. 1–10.
Penneman, R. A., Ryan, R. R., Rosenzweig, A.: Structural Systematics in Actinide Fluoride Complexes. Vol. 13, pp. 1–52.
Politzer, P., Murray, J. S., Grice, M. E.: Charge Capacities and Shell Structures of Atoms. Vol. 80, pp. 101–114.
Powell, R. C., Blasse, G.: Energy Transfer in Concentrated Systems. Vol. 42, pp. 43–96.

Que, Jr., L.: Non-Heme Iron Dioxygenases. Structure and Mechanism. Vol. 40, pp. 39–72.

Ramakrishna, V. V., Patil, S. K.: Synergic Extraction of Actinides. Vol. 56, pp. 35–90.
Raymond, K. N., Smith, W. L.: Actinide-Specific Sequestering Agents and Decontamination Applications. Vol. 43, pp. 159–186.
Reedijk, J., Fichtinger-Schepman, A. M. J., Oosterom, A. T. van, Putte, P. van de: Platinum Amine Coordination Compounds as Anti-Tumour Drugs. Molecular Aspects of the Mechanism of Action. Vol. 67, pp. 53–89.
Reinen, D.: Ligand-Field Spectroscopy and Chemical Bonding in Cr^{3+}-Containing Oxidic Solids. Vol. 6, pp. 30–51.
Reinen, D.: Kationenverteilung zweiwertiger $3d^n$-Ionen in oxidischen Spinell-, Granat- und anderen Strukturen. Vol. 7, pp. 114–154.
Reinen, D., Friebel, C.: Local and Cooperative Jahn-Teller Interactions in Model Structures. Spectroscopic and Structural Evidence. Vol. 37, pp. 1–60.
Reisfeld, R.: Spectra and Energy Transfer of Rare Earths in Inorganic Glasses. Vol. 13, pp. 53–98.
Reisfeld, R.: Radiative and Non-Radiative Transitions of Rare Earth Ions in Glasses. Vol. 22, pp. 123–175.
Reisfeld, R.: Excited States and Energy Transfer from Donor Cations to Rare Earths in the Condensed Phase. Vol. 30, pp. 65–97.
Reisfeld, R., Jørgensen, C. K.: Luminescent Solar Concentrators for Energy Conversion. Vol. 49, pp. 1–36.
Reisfeld, R., Jørgensen, C. K.: Excited States of Chromium(III) in Translucent Glass-Ceramics as Prospective Laser Materials. Vol. 69, pp. 63–96.
Reisfeld, R., Jørgensen, Ch. K.: Optical Properties of Colorants or Luminescent Species in Sol–Gel Glasses. Vol. 77, pp. 207–256.
Russo, V. E. A., Galland, P.: Sensory Physiology of *Phycomyces Blakesleeanus*. Vol. 41, pp. 71–110.
Rüdiger, W.: Phytochrome, a Light Receptor of Plant Photomorphogenesis. Vol. 40, pp. 101–140.
Ryan, R. R., Kubas, G. J., Moody, D. C., Eller, P. G.: Structure and Bonding of Transition Metal-Sulfur Dioxide Complexes. Vol. 46, pp. 47–100.

Sadler, P. J.: The Biological Chemistry of Gold: A Metallo-Drug and Heavy-Atom Label with Variable Valency. Vol. 29, pp. 171–214.

Sakka, S., Yoko, T.: Sol–Gel-Derived Coating Films and Applications. Vol. 77, pp. 89–118.

Schäffer, C. E.: A Perturbation Representation of Weak Covalent Bonding. Vol. 5, pp. 68–95.

Schäffer, C. E.: Two Symmetry Parameterizations of the Angular-Overlap Model of the Ligand-Field. Relation to the Crystal-Field Model. Vol. 14, pp. 69–110.

Scheidt, W. R., Lee, Y. J.: Recent Advances in the Stereochemistry of Metallotetrapyrroles. Vol. 64, pp. 1–70.

Schmid, G.: Developments in Transition Metal Cluster Chemistry. The Way to Large Clusters. Vol. 62, pp. 51–85.

Schmidt, P. C.: Electronic Structure of Intermetallic B 32 Type Zintl Phases. Vol. 65, pp. 91–133.

Schmidt, H.: Thin Films, the Chemical Processing up to Gelation. Vol. 77, pp. 115–152.

Schmidtke, H.-H., Degen, J.: A Dynamic Ligand Field Theory for Vibronic Structures Rationalizing Electronic Spectra of Transition Metal Complex Compounds. Vol. 71, pp. 99–124.

Schneider, W.: Kinetics and Mechanism of Metalloporphyrin Formation. Vol. 23, pp. 123–166.

Schubert, K.: The Two-Correlations Model, a Valence Model for Metallic Phases. Vol. 33, pp. 139–177.

Schultz, H., Lehmann, H., Rein, M., Hanack, M.: Phthalocyaninatometal and Related Complexes with Special Electrical and Optical Properties. Vol. 74, pp. 41–146.

Schutte, C. J. H.: The Ab-Initio Calculation of Molecular Vibrational Frequencies and Force Constants. Vol. 9, pp. 213–263.

Schweiger, A.: Electron Nuclear Double Resonance of Transition Metal Complexes with Organic Ligands. Vol. 51, pp. 1–122.

Sen, K. D., Böhm, M. C., Schmidt, P. C.: Electronegativity of Atoms and Molecular Fragments. Vol. 66, pp. 99–123.

Sen, K.: Isoelectronic Changes in Energy, Electronegativity, and Hardness in Atoms via the Calculations of $\langle r^{-1} \rangle$. Vol. 80, pp. 87–100.

Shamir, J.: Polyhalogen Cations. Vol. 37, pp. 141–210.

Shannon, R. D., Vincent, H.: Relationship Between Covalency, Interatomic Distances, and Magnetic Properties in Halides and Chalcogenides. Vol. 19, pp. 1–43.

Shriver, D. F.: The Ambident Nature of Cyanide. Vol. 1, pp. 32–58.

Siegel, F. L.: Calcium-Binding Proteins. Vol. 17, pp. 221–268.

Simon, A.: Structure and Bonding with Alkali Metal Suboxides. Vol. 36, pp. 81–127.

Simon, W., Morf, W. E., Meier, P. Ch.: Specificity of Alkali and Alkaline Earth Cations of Synthetic and Natural Organic Complexing Agents in Membranes. Vol. 16, pp. 113–160.

Simonetta, M., Gavezzotti, A.: Extended Hückel Investigation of Reaction Mechanisms. Vol. 27, pp. 1–43.

Sinha, S. P.: Structure and Bonding in Highly Coordinated Lanthanide Complexes. Vol. 25, pp. 67–147.

Sinha, S. P.: A Systematic Correlation of the Properties of the f-Transition Metal Ions. Vol. 30, pp. 1–64.

Schmidt, W.: Physiological Bluelight Reception. Vol. 41, pp. 1–44.

Smith D. W.: Ligand Field Splittings in Copper(II) Compounds. Vol. 12, pp. 49–112.

Smith D. W., Williams, R. J. P.: The Spectra of Ferric Haems and Haemoproteins, Vol. 7, pp. 1–45.

Smith, D. W.: Applications of the Angular Overlap Model. Vol. 35, pp. 87–118.

Solomon, E. I., Penfield, K. W., Wilcox, D. E.: Active Sites in Copper Proteins. An Electric Structure Overview. Vol. 53, pp. 1–56.

Somorjai, G. A., Van Hove, M. A.: Adsorbed Monolayers on Solid Surfaces. Vol. 38, pp. 1–140.

Speakman, J. C.: Acid Salts of Carboxylic Acids, Crystals with some "Very Short" Hydrogen Bonds. Vol. 12, pp. 141–199.

Spiro, G., Saltman, P.: Polynuclear Complexes of Iron and Their Biological Implications. Vol. 6, pp. 116–156.

Strohmeier, W.: Problem und Modell der homogenen Katalyse. Vol. 5, pp. 96–117.

Sugiura, Y., Nomoto, K.: Phytosiderophores – Structures and Properties of Mugineic Acids and Their Metal Complexes. Vol. 58, pp. 107–135.

Sykes, A. G.: Plastocyanin and the Blue Copper Proteins. Vol. 75, pp. 175–224.

Tam, S.-C., Williams, R. J. P.: Electrostatics and Biological Systems. Vol. 63, pp. 103–151.

Teller, R., Bau, R. G.: Crystallographic Studies of Transition Metal Hydride Complexes. Vol. 44, pp. 1–82.

Therien, M. J., Chang, J., Raphael, A. L., Bowler, B. E., Gray, H. B.: Long-Range Electron Transfer in Metalloproteins. Vol. 75, pp. 109–130.

Thompson, D. W.: Structure and Bonding in Inorganic Derivatives of β-Diketones. Vol. 9, pp. 27–47.

Thomson, A. J., Williams, R. J. P., Reslova, S.: The Chemistry of Complexes Related to cis-Pt(NH₃)₂Cl₂. An Anti-Tumor Drug. Vol. 11, pp. 1–46.

Tofield, B. C.: The Study of Covalency by Magnetic Neutron Scattering. Vol. 21, pp. 1–87.

Thiel, R. C., Benfield, R. E., Zanoni, R., Smit, H. H. A., Dirken, M. W.: The Physical Properties of the Metal Cluster Compound Au₅₅(PPh₃)₁₂Cl₆. Vol. 81, pp. 1–40.

Trautwein, A.: Mössbauer-Spectroscopy on Heme Proteins. Vol. 20, pp. 101–167.

Tressaud, A., Dance, J.-M.: Relationships Between Structure and Low-Dimensional Magnetism in Fluorides. Vol. 52, pp. 87–146.

Trautwein, A. X., Bill, E., Bominaar, E. L., Winkler, H.: Iron-Containing Proteins and Related Analogs–Complementary Mössbauer, EPR and Magnetic Susceptibility Studies. Vol. 78, pp. 1–96.

Tributsch, H.: Photoelectrochemical Energy Conversion Involving Transition Metal d-States and Intercalation of Layer Compounds. Vol. 49, pp. 127–175.

Truter, M. R.: Structures of Organic Complexes with Alkali Metal Ions. Vol. 16, pp. 71–111.

Umezawa, H., Takita, T.: The Bleomycins: Antitumor Copper-Binding Antibiotics. Vol. 40, pp. 73–99.

Vahrenkamp, H.: Recent Results in the Chemistry of Transition Metal Clusters with Organic Ligands. Vol. 32, pp. 1–56.

Valach, F., Koreň, B., Sivý, P., Melník, M.: Crystal Structure Non-Rigidity of Central Atoms for Mn(II), Fe(II), Fe(III), Co(II), Co(III), Ni(II), Cu(II) and Zn(II) Complexes. Vol. 55, pp. 101–151.

Wallace, W. E., Sankar, S. G., Rao, V. U. S.: Field Effects in Rare-Earth Intermetallic Compounds. Vol. 33, pp. 1–55.

Warren, K. D.: Ligand Field Theory of Metal Sandwich Complexes. Vol. 27, pp. 45–159.

Warren, K. D.: Ligand Field Theory of f-Orbital Sandwich Complexes. Vol. 33, pp. 97–137.

Warren, K. D.: Calculations of the Jahn-Teller Coupling Constants for d^x Systems in Octahedral Symmetry via the Angular Overlap Model. Vol. 57, pp. 119–145.

Watson, R. E., Perlman, M. L.: X-Ray Photoelectron Spectroscopy. Application to Metals and Alloys. Vol. 24, pp. 83–132.

Weakley, T. J. R.: Some Aspects of the Heteropolymolybdates and Heteropolytungstates. Vol. 18, pp. 131–176.

Wendin, G.: Breakdown of the One-Electron Pictures in Photoelectron Spectra. Vol. 45, pp. 1–130.

Weissbluth, M.: The Physics of Hemoglobin. Vol. 2, pp. 1–125.

Weser, U.: Chemistry and Structure of some Borate Polyol Compounds. Vol. 2, pp. 160–180.

Weser, U.: Reaction of some Transition Metals with Nucleic Acids and Their Constituents. Vol. 5, pp. 41–67.

Weser, U.: Structural Aspects and Biochemical Function of Erythrocuprein. Vol. 17, pp. 1–65.

Weser, U.: Redox Reactions of Sulphur-Containing Amino-Acid Residues in Proteins and Metalloproteins, an XPS-Study. Vol. 61, pp. 145–160.

West, D.X., Padhye, S.B., Sonawane, P.B.: Structural and Physical Correlations in the Biological Properties of Transitions Metal Heterocyclic Thiosemicarbazone and S-alkyldithiocarbazate Complexes. Vol. 76, pp. 1–50.

Willemse, J., Cras, J. A., Steggerda, J. J., Keijzers, C. P.: Dithiocarbamates of Transition Group Elements in "Unusual" Oxidation State. Vol. 28, pp. 83–126.

Williams, R. J. P.: The Chemistry of Lanthanide Ions in Solution and in Biological Systems. Vol. 50, pp. 79–119.

Williams, R. J. P., Hale, J. D.: The Classification of Acceptors and Donors in Inorganic Reactions. Vol. 1, pp. 249–281.

Williams, R. J. P., Hale, J. D.: Professor Sir Ronald Nyholm. Vol. 15, pp. 1 and 2.

Wilson, J. A.: A Generalized Configuration-Dependent Band Model for Lanthanide Compounds and Conditions for Interconfiguration Fluctuations. Vol. 32, pp. 57–91.

Winkler, R.: Kinetics and Mechanism of Alkali Ion Complex Formation in Solution. Vol. 10, pp. 1–24.

Wood, J. M., Brown, D. G.: The Chemistry of Vitamin B_{12}-Enzymes. Vol. 11, pp. 47–105.

Woolley, R. G.: Natural Optical Activity and the Molecular Hypothesis. Vol. 52, pp. 1–35.

Wüthrich, K.: Structural Studies of Hemes and Hemoproteins by Nuclear Magnetic Resonance Spectroscopy. Vol. 8, pp. 53–121.

Xavier, A. V., Moura, J. J. G., Moura, I.: Novel Structures in Iron-Sulfur Proteins. Vol. 43, pp. 187–213.

Zanello, P.: Stereochemical Aspects Associated with the Redox Behaviour of Heterometal Carbonyl Clusters. Vol. 79, pp. 101–214.

Zumft, W. G.: The Molecular Basis of Biological Dinitrogen Fixation. Vol. 29, pp. 1–65.